Micro and Nanofabrication Using Self-Assembled Biological Nanostructures

Micro and Nanofabrication Using Self-Assembled Biological Nanostructures

Edited by

Jaime Castillo-León
Winnie E. Svendsen

ELSEVIER

AMSTERDAM • BOSTON • HEIDELBERG
LONDON • NEW YORK • OXFORD • PARIS
SAN DIEGO • SAN FRANCISCO • SINGAPORE
SYDNEY • TOKYO

William Andrew is an imprint of Elsevier

William Andrew is an imprint of Elsevier
The Boulevard, Langford Lane, Kidlington, Oxford, OX5 1GB, UK
225 Wyman Street, Waltham, MA 02451, USA

Notices
Knowledge and best practice in this field are constantly changing. As new research and experience broaden our understanding, changes in research methods, professional practices, or medical treatment may become necessary.

Practitioners and researchers must always rely on their own experience and knowledge in evaluating and using any information, methods, compounds, or experiments described herein. In using such information or methods they should be mindful of their own safety and the safety of others, including parties for whom they have a professional responsibility.

To the fullest extent of the law, neither the Publisher nor the authors, contributors, or editors, assume any liability for any injury and/or damage to persons or property as a matter of products liability, negligence or otherwise, or from any use or operation of any methods, products, instructions, or ideas contained in the material herein.

British Library Cataloguing-in-Publication Data
A catalogue record for this book is available from the British Library.

Library of Congress Cataloging-in-Publication Data
A catalog record for this book is available from the Library of Congress.

ISBN: 978-0-323-29642-7

For information on all William Andrew publications
visit our website at http://store.elsevier.com/

This book has been manufactured using Print On Demand technology. Each copy is produced to order and is limited to black ink. The online version of this book will show color figures where appropriate.

TABLE OF CONTENTS

**Chapter 4   Self-Assembled Peptide Nanostructures
                 for Regenerative Medicine and Biology 63**

Ming Ni, Charlotte A.E. Hauser

**Chapter 5   Fabrication of Drug Delivery Systems Using
                 Self-Assembled Peptide Nanostructures 91**

Daniel Keith, Honggang Cui

CONTRIBUTORS

Colin J. Barrow
Center for Chemistry and Biotechnology, Deaking University, Victoria, Australia

Jaime Castillo-León
DTU Nanotech, Kgs. Lyngby, Technical University of Denmark, Lund, Sweden

Honggang Cui
Department of Chemical and Biomolecular Engineering and Institute for NanoBioTechnology, The Johns Hopkins University, Baltimore, Maryland

Juliet A. Gerrard
MacDiarmid Institute for Advanced Materials and Nanotechnology, New Zealand

Charlotte A.E. Hauser
Institute of Bioengineering and Nanotechnology, The Nanos, Singapore, Singapore

Daniel Keith
Department of Chemical and Biomolecular Engineering and Institute for NanoBioTechnology, The Johns Hopkins University, Baltimore, Maryland

Rui Li
Center for Chemistry and Biotechnology, Deakin University, Victoria, Australia

Ming Ni
Institute of Bioengineering and Nanotechnology, The Nanos, Singapore, Singapore

David R. Nisbet
Research School of Engineering, The Australian National University, Canberra, Acton, Australia

Alexandra Rodriguez
Research School of Engineering, The Australian National University, Canberra, Acton, Australia

Luigi Sasso
Biomolecular Interaction Centre and School of Biological Sciences, University of Canterbury, New Zealand

Richard J. Williams
Center for Chemistry and Biotechnology, Deakin University, Victoria, Australia

Micro and Nanofabrication Using Self-Assembled Biological Nanostructures is a result of the research activities of a group of scientists worldwide who wrote the chapters. This book is dealing with the possibilities and challenges when using self-assembled peptide nanostructures for the fabrication of cell scaffolds, drug-delivery systems, biosensing platforms, and new nanostructures. In a very didactic way this work presents the state of the art in the use of self-assembled biological nanostructures, and its advantages and challenges.

We would like to thank all the authors for their effort in helping us to put together this work. We also would like to thank Elsevier for inviting us to publish this book and all the support during its preparation.

We hope that this book will motivate a broad audience of students and researchers interested in self-assembled biological nanostructures so as to consider these technologies as an alternative in their research resulting in new and exciting applications.

Jaime Castillo-León
Winnie E. Svendsen
2014

CHAPTER 1

Self-Assembled Biological Nanofibers for Biosensor Applications

Luigi Sasso[1,2], Juliet A. Gerrard[1,2]
[1]Biomolecular Interaction Centre and School of Biological Sciences, University of Canterbury, New Zealand
[2]MacDiarmid Institute for Advanced Materials and Nanotechnology, New Zealand

1.1 INTRODUCTION

The use of nanomaterials in analytical systems has been gradually growing, due to the promise of increased precision and sensitivity offered by the nanoscale dimensions. When used as nanoscaffolds, the high surface-to-volume ratios provided by nanomaterials offer an increase in immobilization of functional sensing components, resulting in the ability to detect lower analyte concentrations [1].

Bionanomaterials, such as biological nanofibers, offer particular advantages over their inorganic counterparts: the laboratory conditions for their self-assembly and synthesis are mild when compared with harsh solvents and environments found in inorganic nanofabrication techniques; there are available primary sources for biological nanofibers originating from low-value coproducts of the pharmaceutical and biotech industry, offering cost-effectiveness and an environmentally friendly approach to the production of nanomaterials [2–5]; the biological nature of bionanofibers adds a biocompatibility element to their applications, as well as a versatility in functionalization due to the variety of chemical moieties provided by the amino-acid residues in their protein and peptide building blocks. Because of these benefits, biological nanofibers have been gathering considerable attention over the past decades by the scientific community, expanding the field by studying this bionanomaterial and showing its potential to be applied not only in biosensor systems, but also in a variety of bionanotechnological applications such as drug delivery systems, bioelectronics, and tissue reparation [6].

Micro and Nanofabrication Using Self-Assembled Biological Nanostructures. DOI: 10.1016/B978-0-323-29642-7.00001-1

Although bionanofibers are a promising material and there are continuous advances in their integration with analytical detection systems, there are still major challenges that have to be overcome before being able to fully commercialize their use in biosensing. Even lab-bench approaches of their use result in issues and limitations – mostly a function of the same characteristics that make this material attractive. Their stability in extreme temperatures, solvents, and conditions, required for their commercial use, renders some types of bionanofibers unusable [7–13]. The variety in surface moieties available for functionalization, one of the most attractive features of bionanofibers, is on the other hand also a hurdle to commercialization, since a specific functionalization strategy is needed for different systems, depending both on the type of bionanofiber, the primary source, and the functional biosensing element [14–18]. The precise manipulation and characterization of bionanofibers is an additional challenge, and novel nanotechnological methods must be developed to achieve this [13,19–23]. Finally, the poor conductivity of bionanomaterials requires additional consideration, as charge-transfer is a major parameter in biosensing systems.

1.2 TYPES OF SELF-ASSEMBLED BIOLOGICAL NANOFIBERS

Via self-assembly, biological building blocks, such as peptides and proteins, can yield a variety of nanofibers, each with specific characteristics reflecting the source material and the self-assembly conditions. Natural proteins tend to form wire-like, flexible nanostructures with high internal structural order, resulting in stability in natural environments. Artificially designed biological building blocks, on the other hand, will self-assemble into a variety of well-defined nanofibers such as nanotubes, nanowires, or nanofibrils, each with specific properties and advantages.

1.2.1 Natural Protein Nanofibers

There are several protein sources that naturally self-assemble into fibrous nanostructures with mechanical and chemical properties that render them useful in biosensors applications. From collagen and actin to cellular microtubules, nature has provided us with a myriad of biological building blocks that, under the right conditions, can be made to self-assemble into bionanofibers *in vitro* [24].

Within this range of naturally occurring bionanofibers, silkworm and spider silk are of particular interest to the scientific community, due to their simple molecular design and the impressive mechanical strengths that these materials exhibit [25–27]. The primary proteins present in silk are fibroin and sericin, although there is a wide variation between silk-producing species, and therefore also in the structural conformation of the silk nanofibers produced [28].

Silk nanofibers have been extensively integrated into nano- and microtechnology, and micropatterned silk proteins have been used to develop biocompatible nanoscale biosensor platforms [29], utilizing a metallization method to create free-standing silk/metal composite nanofibers [30]. The photonic properties of silk nanofibers have also been exploited for the development of optical biosensor platforms [31,32], and the specific biocompatibility of this nanomaterial has moved the field toward implantable biosensor systems [33].

1.2.2 Amyloid Fibrils

Amyloid fibrils are insoluble nanostructures resulting from the self-assembly of unfolded protein monomers. They hold a distinctive quaternary structure predominantly characterized by a rich, hydrogen-bonded, β-sheet conformation [34,35], a configuration with a rigid internal order that provides nanostructures with high strength, stability, and high-morphological aspect ratios [12,36]. It is perhaps their stability and insolubility in aqueous media that renders amyloid fibrils favorable in biosensor applications [17,37–39] with advantages over other biological nanofibers such as peptide nanotubes [11].

Because of the versatility of protein monomers as molecular building blocks, the source proteins that have been shown to undergo fibrillation under controlled environmental conditions are abundant and span a wide range of biological materials. The list includes insulin [15,40–42], considered a historical standard amyloid fibril source for bionanotechnological applications, fungal hydrophobins [43], ovalbumin [2], and lysozyme [44]. Important additions to this list have been proteins that are considered industrial waste materials, such as whey proteins [4,5,45–48] and eye lens crystallins [3,36,49].

1.2.3 Peptide Nanotubes and Nanowires

Inspired by Nature's great success in forming self-assembled fibrous materials, researchers have synthesized peptide monomers able to recreate protein-like interactions, by focusing on the same amino-acid repeat units that play a key role in the self-assembly of proteins [50,51].

One peptide in particular has attracted much attention from researchers in the field over the past decade: diphenylalanine. This aromatic dipeptide is known for being the core recognition motif of the Alzheimer's disease β-amyloid polypeptide, and its use in bionanotechnology has been encouraged by the mild environment (room temperature, aqueous conditions) needed for its self-assembly. An additional advantage of using peptides as molecular building blocks is their ability to form several different nanostructures depending on the environmental conditions used during the self-assembly [52], such as nanotubes [53,54] and nanowires [55–57].

Peptide nanotubes, hollow nanofiber structures formed by, among others, the diphenylalanine peptide, have been extensively considered for the application of self-assembled nanostructures in biosensor platforms. The amyloid-like aromatic stacking that allows the self-assembly of peptide nanotubes confers an unusual strength to this nanomaterial [58,59]. This property, along with an ease in functionalization offered by the chemistry available on their surface, has allowed for the creation of several biosensor platforms based on the use of peptide nanotubes [60–68].

Diphenylalanine nanowires, solid rod-like bionanofibers, have been gaining scientific interest because they offer most of the positive attributes of their tubular nanostructure counterparts, but with an additional chemical and thermal stability [10] that the nanotubes do not possess [11]. The nanowires' surface-dependent growth mechanism is also one of their advantages, especially for surface-based biosensor platforms. Although their synthesis involves aniline vapor and does require higher temperatures than for nanotube the self-assembly [55,56], covering a surface with peptide nanowires creates a biocompatibility useful especially for biosensor platforms tailored toward biological cell analysis [69–73] (Table 1.1).

Table 1.1 Examples of the Variety of Biological Nanofibers That Have Been Used in Biosensor Platforms

Bionanofiber	Source Biomolecule	Sensor Target	Biosensing Technique	References
Silk nanofibers	Silk proteins	Refractive index	Optical	[31,32]
Amyloid fibrils	Whey proteins	Glucose	Electrochemical	[17]
	β-Lactoglobulin	Humidity, enzyme activity	Mechanical	[37]
	β-Lactoglobulin	Humidity, pH	Electrical	[38]
Peptide nanotubes	Diphenylalanine	Glucose, NADH, H_2O_2, and ethanol	Electrochemical	[60–63]
	Naphthylalanine-naphthylalanine	Phenols	Electrochemical	[62]
	Cyclic peptides	*Escherichia coli*	Electrochemical	[66]
	Bola-amphiphilic peptides	Pathogens	Electrochemical	[67]
Peptide nanowires	Diphenylalanine	H_2O_2	Electrochemical	[69,72]
	Ionic-complementary peptides	Glucose	Electrochemical	[65]

1.3 PRACTICAL LABORATORY CONSIDERATIONS

The mild conditions for self-assembly vary depending on the protein or peptide source used as building block, and some systems require specific pH or temperature conditions during the assembly, and a different set of conditions for stability and storage. The effect of postassembly conditions on the nanofibers' structural integrity and stability are especially important when utilizing the nanofibers in biosensor platforms that require extreme conditions such as high voltages, temperatures, or extreme pHs.

1.3.1 Temperature Stability

Amyloid fibrils are known for their stability, which includes a resistance to thermal stress up to 100°C [74], due primarily by their highly ordered internal structure [75]. It is important to note that when working with amyloid fibrils, a temperature-related instability occurs at the lower temperature extremes instead of the higher ones. Domigan et al. have investigated the effect of storage temperatures on the morphology of amyloid fibrils prepared from bovine insulin [36]. The fibrils' morphology was unchanged upon long-time storage at room temperature, but subjecting amyloid fibrils to freezing/thawing cycles resulted in fibril fracturing, yielding very short fibrils with lengths in the range of 1/10 of their original dimensions. Although the amyloid fibril structure remained intact, validated by fluorescence-based techniques, the integrity of the nanofibrils' morphological properties was lost. This important effect needs to be taken into consideration when utilizing amyloid fibrils in biosensor platforms, especially those requiring biocompatibility, since fibril fragmentation is known for increasing cytotoxicity [76]. Additionally, the freezing/thawing effect on fibril fragmentation is important to consider when shipping the material via airfreight, where the material is subjected to extreme low temperatures.

Peptide nanotubes, although used extensively in electrochemical biosensor platforms, encounter some issues in stability when compared with other biological nanofibers. High temperatures have been found to affect the peptide nanotubes' structural stability, and structural degradation has been observed when diphenylalanine nanotubes have been exposed to temperatures higher than 100–150°C [8,9,21]. A phase change has been observed in this temperature range, corresponding to a decrease in polarization and a corresponding phase transition at ~140°C [77]. This degradation at high temperatures is particularly cumbersome when wanting to utilize these nanofibers in biosensor platforms that require autoclave-based sterilization. As an alternative, peptide nanowires have proven to be much more robust in terms of temperature stability. Their self-assembly already requires temperatures above 100°C [55,56], and the nanofibers can withstand temperatures as high as 200°C, with a thermal decomposition only observed at ~250–300°C [10].

1.3.2 Solvents and Solubility

The stability of biological nanofibers in liquid environments is essential to their application in biosensor platforms, especially in systems where the nanofibers act as nanoscaffolds for a functional component such as enzymes or other active biomolecules. Degradation of the nanoscaffolds can result in a leakage of those components, and ultimately in an instability and lack of reproducibility of sensorial parameters.

Although diphenylalanine peptide nanotubes have been used to create proof-of-principle biosensor platforms, these nanofibers have been found at a later date to solubilize in aqueous solutions and methanol [11]. This limitation renders them unemployable in liquid-based systems, and research has moved toward utilizing other insoluble nanofibers for this purpose, like peptide nanowires from the same source dipeptide [10] or amyloid fibrils, which are insoluble by definition and have been specifically tested for solubility in common aqueous biological solvents and solvents used in microfabrication processes such as ethanol, isopropanol, methanol, and acetonitrile (with only the last one having a severe degradation effect onto the nanofibrils) [13].

1.4 FUNCTIONALIZATION APPROACHES

The majority of applications of biological nanofibers in biosensor platforms involve the use of the nanostructures as scaffolds for sensing components such as enzymes, antibodies, or other relevant biomolecules. The attachment of those components onto the nanofibers can rely on several mechanisms and interactions (see Figure 1.1), and is mainly dependent on the activity and stability of the functional components.

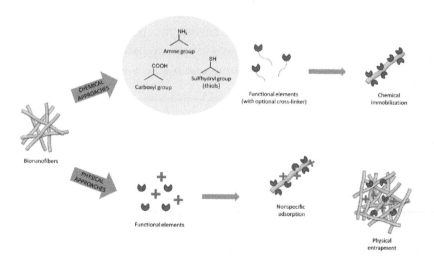

Fig. 1.1. Schematic of common bionanofiber functionalization approaches.

1.4.1 Chemical Approaches

Chemistry-based approaches are preferred for nanofiber functionalization because of the many advantages they offer, including stronger nanofiber/biomolecule attachment and long-term system stability. Additionally, the variety of chemical moieties present on the nanofiber surface offers a favorable approach in exploiting protein chemistry and other biological interactions, using the amino acid residues on the surface of the functional components (e.g., enzymes and antibodies).

Cross-linking is a functionalization approach based on strong covalent bonds, and for this reason is considered the most stable immobilization method [78]. Cross-linker molecules can act as a bridge in functional component/nanofiber interactions with free moieties that are not necessarily compatible. This functionalization mechanism has alternative advantages, due to the flexibility in cross-linking conditions, spacer arm, and specificity in chemical orientation of the functional component. This latter concept is extremely important in biosensing systems that require a specific activity of the functional component, such as enzymatic biosensing mechanisms, where a loss of enzymatic activity can occur due to conformational restrictions [79].

The active functional groups that are primary targets for the functionalization are primary amines (present in lysine residues) [14,17], thiols (cysteine residues) [80], and carboxylic acids (aspartic and glutamic acid residues) [65,66]. These groups are commonly used in protein chemistry, and their operational chemistry is well understood and characterized.

1.4.2 Physical Approaches

For cases where a specific functionalization is not possible, either because of poor nanofiber surface chemistry, lack of specificity of cross-linker interactions, or high costs of functionalization reagents, other nonspecific functionalization approaches can be used.

Physical adsorption is the most cost-effective method for nanofiber functionalization, and its simplicity is an attractive advantage over more specific techniques. Functional components can be attached to a nanofiber scaffold by simple physical interactions such as hydrogen bonding, Van der Waals forces, or ionic attractions [81–83]. Adsorption can be achieved in mild conditions, and as a noninvasive method it rarely affects the functionality of the immobilized component, e.g., enzymatic activity. Being nonspecific, adsorption also allows for a facile multifunctionalization of the nanofibers. There are issues though with this method, and often the attachment is weak and can result in functional component dissociating and therefore leakage over long periods of time or under changes in environmental conditions such as ionic strength, temperature, or pH [84].

An additional nonspecific, but stronger and more stable approach is network physical entrapment. When used in relatively high concentrations, biological nanofibers can act as a polymeric network that, when assembled onto a surface, can encapsulate functional components. The greatest disadvantage of physical entrapment is the mass transfer limitations that can arise depending on the porosity of the obtained nanofiber networks. An optimization of entrapment parameters should be carried out to allow analytes to pass through the network while retaining the encapsulated functional components to avoid their leakage (Table 1.2).

1.5 COMMON CHALLENGES IN BIOSENSOR PLATFORMS

The major factors that are limiting the application of bionanofibers in biosensor platforms are caused by the small dimensions of the individual nanofibers as well as their biological nature. The individual manipulation of biological nanostructures is crucial when developing biosensor platforms that require single nanofibers to be placed in a specific location (see Figure 1.2). Fortunately, nanotechnological tools are in constant development, and there are several techniques that can be used for achieving precise handling of bionanofibers. An additional common challenge in bionanofiber-based biosensor development is the poor electrical properties of biological materials, a trait that must be kept under consideration especially in electronic and electrochemical detection systems. Mindful consideration of these attributes, though, can aid in optimizing conditions to ultimately overcome these charge-transfer limitations.

Table 1.2 Examples of Postassembly Functionalization Approaches for Biological Nanofibers

Bionanofiber	Source Biomolecule	Functional Component	Functionalization Approach	References
Amyloid fibrils	Insulin	Glucose oxidase	Cross-linking (glutaraldehyde)	[14]
	Insulin	Pt complexes, PEDOT	Ionic interactions	[15,83]
	Whey proteins	Glucose oxidase	Cross-linking (biotinylation)	[17]
	β-Lactoglobulin	Retinoic acid, discotic protoporphyrine, carboxyfullerene ligands	Ligand-doped coassembly	[16]
	β-Lactoglobulin	TiO_2 nanoparticles, PEDOT	Ionic interactions	[81]
	Recombinant silk proteins	PEDOT	Ionic interactions	[82]
Peptide nanotubes	Diphenylalanine	Glucose oxidase	Cross-linking (glutaraldehyde)	[60,61]
	Diphenylalanine, naphthylalanine-naphthylalanine	Tyrosinase	Physical entrapment	[62]
	Diphenylalanine	Microperoxidase	Layer-by-layer deposition	[63]
	Ionic-complementary peptides	Glucose oxidase	Cross-linking (amide bonds)	[65]
	Cyclic peptides	Antibodies	Cross-linking (amide bonds)	[66]
	Bola-amphiphilic peptides	Antibodies	Physical adsorption	[67]
	Diphenylalanine	Glucose oxidase and horseradish peroxidase	Physical encapsulation	[68]
Peptide nanowires	Leucine zipper motif	NADH peroxidase	Cross-linking (disulfide bonds)	[80]

1.5.1 Manipulation

Very often the use of biological nanofibers in biosensor systems requires their placement onto a surface or specified transducer-acting location. For this reason, a controlled nanofiber manipulation is crucial, and the manipulation technique chosen must require environmental conditions that do not damage the biomaterial or its functional components. Advances in micro- and nanotechnology have provided tools for handling bionanofiber samples [6,20,23,85], but choosing the right technique for handling sticky, nanosized fibers without damaging them can be a

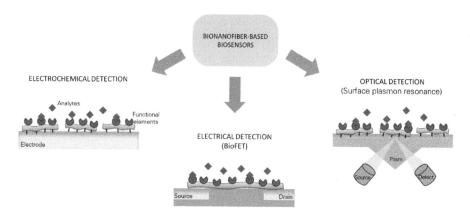

Fig. 1.2. Schematic of typical bionanofiber-based biosensor platforms.

challenge. The following are some possible and most commonly used manipulation techniques useful in biosensor platforms.

1.5.1.1 AFM-Based Techniques

Because of the small dimensions of the nanofibers, manipulation techniques with handling components of the same micro- and nano-size dimensional ranges are often necessary. AFM instruments involve microtips with those exact dimensions, providing a useful tool to push, pull, or drag biological nanofibers. An enormous advantage of AFM-based manipulation techniques is their ability to work in a liquid environment, an instrument property especially useful when working with biological nanostructures. This direct-contact tool can be used to pattern amyloid nanofibrils on a surface by creating charge patterns then used to attract the nanofibers [86], by creating nucleation sites on a peptide filament [22], and by nanolithography of metal/peptide nanofiber complexes [87]. This technique has proven to be especially useful in surface-based biosensor platforms such as electrochemical systems, where the peptide fibers act as nanoscaffolds for redox-active enzymes [65,88].

1.5.1.2 Dielectrophoresis

Dielectrophoresis (DEP) is one of the least invasive techniques for the manipulation of biological samples [20,85]. The principle behind this method is the nanostructure polarization achieved by the presence of an inhomogeneous electric field, while in solution. The charges accumulated around the nanofiber will either attract it toward or repel it from the electrodes that are generating the electric field. When the electrodes are spaced apart with a distance smaller than the nanofiber length, it is possible to obtain a system where a biological nanofiber is immobilized between two individually addressable electrodes, therefore creating a field-effect transistor systems [13,67,71]. Such a system is both useful for an investigation of the nanofibers' conductive properties and in the creation of a biosensing platform, if the nanofibers are functionalized with an analyte-binding component that will generate a change in capacitance upon binding [67].

1.5.1.3 Soft-Lithography

Soft-lithography is a standard microfabrication technique that can be utilized for patterning biological nanofibers. This method is particularly useful for surface-based biosensor platforms, which require a deposition or attachment of bionanofibers onto specific locations. This method is based on the creation of surface patterns with a sacrificial mold that is removed after the bionanofiber deposition.

Soft polymers such as polydimethylsiloxane (PDMS) have been used to create molds with micron-sized channels, placed onto a silicon surface. By allowing a peptide solution to fill up the channels by capillary forces, and then by removing the mold and tailoring the conditions for peptide nanowire growth, patterned arrays of diphenylalanine peptide nanowires have been created onto silicon surfaces [56]. This technique has been applied onto interdigitated electrode arrays to create an electrochemical biosensing platform with increased biocompatibility provided by the peptide nanowires [72].

A similar approach has been used for the patterning of nanotubes from the same peptide, using sacrificial molds based on aluminum foils [54] and silicon dioxide layers [89].

1.5.1.4 Chemical Immobilization

Alternatively, the functionalization approaches discussed previously can be utilized to immobilize the bionanofibers onto biosensor platforms via their surface moieties and their chemical interactions with a surface of a specific material. In the case of gold metal electrodes, for example, successful immobilizations have been achieved via direct thiolation of the bionanofiber surfaces [17,60–62], or via a functionalization of the electrode surface for bionanofiber attachment [68].

1.5.2 Conductive Properties in Bio-FETs

Field effect transistor (FET) sensors utilize the intrinsic electric properties of a sensorial transistor to measure changes in conductivity due to chemical changes on its surface. Reducing the size of the sensorial component to nanoscale dimensions increases its surface-to-volume ratio and therefore the electrical sensitivity of the system. Often, the sensorial component of FET sensors is a single semiconductor nanostructure, such as a carbon nanotube [90] or a silicon nanowire [91], able to carry currents with high-precision sensitivities.

Recent advances in bionanofiber manipulation and characterization techniques have led to the inclusion of biological nanofibers in the development of novel bio-FET sensor platforms. Such platforms can be designed so that the bionanofibers act as either the sensorial component, i.e., a "conductive biological nanowire," or as a support substrate to include additional advantages to the platform.

As mentioned previously, DEP forces can be used to place single or bundles of bionanofibers in between microelectrodes, therefore creating a bio-FET platform where the immobilized bionanofiber is intended as the sensorial component. This technique has been used for the assembly of bio-FET platforms in tailor-made microelectrode platforms using peptide nanotubes [19,67] and amyloid fibrils [13,36]. Interdigitated electrodes have also been used to create platforms with elastin-related polypeptide nanofibrils [92]. Although the development of a fully working bio-FET that utilizes bionanofibers as the functional component is still ongoing optimization within the scientific community, these platforms and system assemblies have been useful in the electrical characterization of the immobilized bionanofibers.

The cited examples utilize different bionanofibers and electrode designs, but they share similar challenges, ultimately related to similar poor conductive properties among the various types of bionanofibers. This effect has been found to be strongly related to the bionanofiber morphology, meaning that controlling the assembly mechanism can lead to a direct control of the bionanofiber's charge transport properties [93].

Alternatively, biological nanofibers can be used as a support material for the fabrication of carbon nanotube-FETs on nanofibrillated cellulose paper [94]. The main advantage of using a nanofibrillated cellulose paper lies with the optical properties related with the small dimensions of the cellulose fibers. Decreasing the dimensions of the paper fibers in fact reduces the optical scattering of the material, yielding a nanopaper support with excellent optical transparency and better thermal stability than other similar substrates [94].

1.5.3 Charge-Blocking Behavior in Electrochemical Platforms

One of the greatest challenges when working with biological nanomaterials is the poor conductive behavior that results from a lack of charge carriers along the nanofibers' surface. Although small currents have been measured passing through bionanofibers (see 1.5.2), these are not of high enough magnitude to compete with the charge transfer processes required in electrochemical biosensing systems.

A typical bionanofiber-based electrochemical sensor utilizes the nanostructures as scaffolds for the immobilization of redox-active functional components, such as enzymes or antibodies, onto the working electrode surface. Optimizing the conditions in order to achieve the necessary sensitivity in current response is a challenging task, and requires careful characterization of the insulating properties, or "blocking behavior," of the nanofibers. The concentration of nanofibers suspension used for the deposition onto an electrode surface is crucial, as it is often directly reflected on the material's physical coverage of the electrode surface. This phenomenon can be easily characterized by voltammetric measurements in electrolyte solutions typically used, such as the ferri/ferrocyanide redox couple, while comparing electrodes with bionanofibers deposited at different concentrations. Results from literature suggest, as expected, that the current response magnitude decreases with increasing bionanofiber concentration [17,69]. On the other hand, when using bionanofibers as scaffolds for functional biosensing components such as enzymes, their deposition on the electrodes must be with high enough concentrations to allow for enough functional components to adhere to the surface.

The overall performance of the biosensor platform must additionally be assessed in terms of sensitivity and detection limit, stability over time in various environmental conditions such as pH and temperature, selectivity and reproducibility.

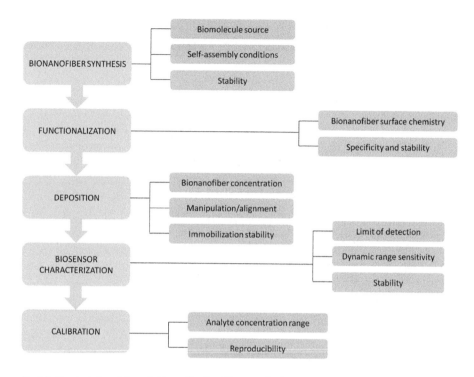

Fig. 1.3. Flowchart of crucial steps in bionanofiber-based biosensor development, with important key parameters for each.

1.6 CONCLUSIONS

This chapter gave an overview of the use of self-assembled biological nanofibers in biosensor systems, highlighting how to overcome the major challenges and issues encountered when working with this specific nanomaterial. Advances in the field have provided us with numerous strategies available for the immobilization of functional components, for selecting the best conditions to achieve stability and reproducibility, and how to best use the nanotechnological techniques at our disposal for the characterization and manipulation of bionanofibers.

The use of biological nanofibers in biosensor systems is a relatively new scientific advancement, and the field is under increasing constant development. Although it might still be too early to obtain stable, reproducible systems ready for commercialization, it is only a matter of time before these technologies will be able to be translated from the lab-bench to the market, providing a cost-effective, environmentally conscious approach to biosensor development.

REFERENCES

[1] Z. Wang, S. Liu, P. Wu, C. Cai, Detection of glucose based on direct electron transfer reaction of glucose oxidase immobilized on highly ordered polyaniline nanotubes, Anal Chem 81 (2009) 1638–1645.

[2] F.G. Pearce, S.H. Mackintosh, J.A. Gerrard, Formation of amyloid-like fibrils by ovalbumin and related proteins under conditions relevant to food processing, J Agric Food Chem 55 (2007) 318–322.

[3] M. Garvey, S.L. Gras, S. Meehan, S.J. Meade, J.A. Carver, J.A. Gerrard, Protein nanofibres of defined morphology prepared from mixtures of crude crystallins, Int J Nanotechnol 6 (2009) 258–273.

[4] S.M. Loveday, J. Su, M.A. Rao, S.G. Anema, H. Singh, Whey protein nanofibrils: kinetic, rheological and morphological effects of group IA and IIA cations, Int Dairy J 26 (2012) 133–140.

[5] S.M. Loveday, J. Su, M.A. Rao, S.G. Anema, H. Singht, Whey protein nanofibrils: the environment-morphology-functionality relationship in lyophilization, rehydration, and seeding, J Agric Food Chem 60 (2012) 5229–5236.

[6] J. Castillo, K.B. Andersen, W.E. Svendsen, Self-assembled peptide nanostructures for biomedical applications: advantages and challenges, in: R. Pignatello (Ed.), Biomaterials science and engineering, InTech, 2011.

[7] O.S. Makin, E. Atkins, P. Sikorski, J. Johansson, L.C. Serpell, Molecular basis for amyloid fibril formation and stability, Proc Natl Acad Sci USA 102 (2005) 315–320.

[8] L. Adler-Abramovich, M. Reches, V.L. Sedman, S. Allen, S.J.B. Tendler, E. Gazit, Thermal and chemical stability of diphenylalanine peptide nanotubes: implications for nanotechnological applications, Langmuir 22 (2006) 1313–1320.

[9] V.L. Sedman, L. Adler-Abramovich, S. Allen, E. Gazit, S.J.B. Tendler, Direct observation of the release of phenylalanine from diphenylalanine nanotubes, J Am Chem Soc 128 (2006) 6903–6908.

[10] J. Ryu, C.B. Park, High stability of self-assembled peptide nanowires against thermal, chemical, and proteolytic attacks, Biotech Bioeng 105 (2010) 221–230.

[11] K.B. Andersen, J. Castillo-Leon, M. Hedstrom, W.E. Svendsen, Stability of diphenylalanine peptide nanotubes in solution, Nanoscale 3 (2011) 994–998.

[12] O.G. Jones, R. Mezzenga, Inhibiting, promoting, and preserving stability of functional protein fibrils, Soft Matter 8 (2012) 876–895.

[13] L. Domigan, K.B. Andersen, L. Sasso, M. Dimaki, W.E. Svendsen, J.A. Gerrard, et al. Dielectrophoretic manipulation and solubility of protein nanofibrils formed from crude crystallins, Electrophoresis 34 (2013) 1105–1112.

[14] S.M. Pilkington, S.J. Roberts, S.J. Meade, J.A. Gerrard, Amyloid fibrils as a nanoscaffold for enzyme immobilization, Biotech Progr 26 (2010) 93–100.

[15] Q. Tang, N. Solin, J. Lu, O. Inganas, Hybrid bioinorganic insulin amyloid fibrils, Chem Comm 46 (2010) 4157–4159.

[16] N. Hendler, B. Belgorodsky, E.D. Mentovich, S. Richter, L. Fadeev, M. Gozin, Multiple self-assembly functional structures based on versatile binding sites of beta-lactoglobulin, Adv Funct Mat 22 (2012) 3765–3776.

[17] L. Sasso, S. Suei, L. Domigan, J. Healy, V. Nock, M.A. Williams, et al. Versatile multi-functionalization of protein nanofibrils for biosensor applications, Nanoscale 6 (2014) 1629–1634.

[18] J.K. Raynes, F.G. Pearce, S.J. Meade, J.A. Gerrard, Immobilization of organophosphate hydrolase on an amyloid fibril nanoscaffold: towards bioremediation and chemical detoxification, Biotech Progr 27 (2011) 360–367.

[19] J. Castillo, S. Tanzi, M. Dimaki, W. Svendsen, Manipulation of self-assembly amyloid peptide nanotubes by dielectrophoresis, Electrophoresis 29 (2008) 5026–5032.

[20] J. Castillo, M. Dimaki, W.E. Svendsen, Manipulation of biological samples using micro and nano techniques, Integr Biol 1 (2009) 30–42.

[21] V.L. Sedman, S. Allen, X.Y. Chen, C.J. Roberts, S.J.B. Tendler, Thermomechanical manipulation of aromatic peptide nanotubes, Langmuir 25 (2009) 7256–7259.

[22] F.C. Zhang, F. Zhang, H.N. Su, H. Li, Y. Zhang, J. Hu, Mechanical manipulation assisted self-assembly to achieve defect repair and guided epitaxial growth of individual peptide nanofilaments, ACS Nano 4 (2010) 5791–5796.

[23] J. Castillo-Leon, L. Sasso, W.E. Svendsen, Manipulation of self-assembled peptide nano-structures, in: J. Castillo-Leon, L. Sasso, W.E. Svendsen (Eds.), Self-assembled peptide nanostructures. Advances and applications in nanobiotechnology, Taylor & Francis, 2012, pp. 125–145.

[24] L.J. Domigan, Proteins and peptides as biological nanowires: towards biosensing devices, in: J.A. Gerrard (Ed.), Protein nanotechnology: protocols, instrumentations and applications, 2nd ed., Humana Press, 2013, pp. 131–152 996.

[25] B. Kundu, N.E. Kurland, S. Bano, C. Patra, F.B. Engel, V.K. Yadavalli, et al. Silk proteins for biomedical applications: bioengineering perspectives, Prog. Polym. Sci. 39 (2014) 251–267.

[26] G.H. Altman, F. Diaz, C. Jakuba, T. Calabro, R.L. Horan, et al. Silk-based biomaterials, Biomaterials 24 (2003) 401–416.

[27] H. Tao, D.L. Kaplan, F.G. Omenetto, Silk materials – a road to sustainable high technology, Adv Mat 24 (2012) 2824–2837.

[28] H.J. Jin, D.L. Kaplan, Mechanism of silk processing in insects and spiders, Nature 424 (2003) 1057–1061.

[29] H. Perry, A. Gopinath, D.L. Kaplan, L. Dal Negro, F.G. Omenetto, Nano- and micropatterning of optically transparent, mechanically robust, biocompatible silk fibroin films, Adv Mat 20 (2008) 3070–3072.

[30] K. Tsioris, H. Tao, M.K. Liu, J.A. Hopwood, D.L. Kaplan, R.D. Averitt, et al. Rapid transfer-based micropatterning and dry etching of silk microstructures, Adv Mat 23 (2011) 2015–2019.

[31] S. Kim, A.N. Mitropoulos, J.D. Spitzberg, D.L. Kaplan, F.G. Omenetto, Silk protein based hybrid photonic-plasmonic crystal, Opt Express 21 (2013) 8897–8903.

[32] D.M. Lin, H. Tao, J. Trevino, J.P. Mondia, D.L. Kaplan, F.G. Omenetto, et al. Direct transfer of subwavelength plasmonic nanostructures on bioactive silk films, Adv Mat 24 (2012) 6088–6093.

[33] H. Tao, J.M. Kainerstorfer, S.M. Siebert, E.M. Pritchard, A. Sassaroli, B.J. Panilaitis, et al. Implantable, multifunctional, bioresorbable optics, Proc Natl Acad Sci USA 109 (2012) 19584–19589.

[34] M. Faendrich, On the structural definition of amyloid fibrils and other polypeptide aggregates, Cell Mol Life Sci 64 (2007) 2066–2078.

[35] F. Chiti, C.M. Dobson, Protein misfolding, functional amyloid, and human disease, Annu Rev Biochem 75 (2006) 333–366.

[36] L.J. Domigan, J.P. Healy, S.J. Meade, R.J. Blaikie, J.A. Gerrard, Controlling the dimensions of amyloid fibrils: toward homogenous components for bionanotechnology, Biopolymers 97 (2012) 123–133.

[37] C. Li, J. Adamcik, R. Mezzenga, Biodegradable nanocomposites of amyloid fibrils and graphene with shape-memory and enzyme-sensing properties, Nat Nanotechnol 7 (2012) 421–427.

[38] C. Li, S. Bolisetty, R. Mezzenga, Hybrid nanocomposites of gold single-crystal platelets and amyloid fibrils with tunable fluorescence, conductivity, and sensing properties, Adv Mat 25 (2013) 3694–3700.

[39] S.L. Gras, Surface- and solution-based assembly of amyloid fibrils for biomedical and nanotechnology applications, in: J.K. Rudy (Ed.), Advances in chemical engineering, vol. 35, Academic Press, 2009, pp. 161–209.

[40] C. Ha, C.B. Park, Template-directed self-assembly and growth of insulin amyloid fibrils, Biotech Bioeng 90 (2005) 848–855.

[41] Q.X. Hua, M.A. Weiss, Mechanism of insulin fibrillation – the structure of insulin under amyloidogenic conditions resembles a protein-folding intermediate, J Biol Chem 279 (2004) 21449–21460.

[42] M. Groenning, S. Frokjaer, B. Vestergaard, Formation mechanism of insulin fibrils and structural aspects of the insulin fibrillation process, Curr Prot Pep Sci 10 (2009) 509–528.

[43] V.K. Morris, M. Sunde, Formation of amphipathic amyloid monolayers from fungal hydrophobin proteins, in: J.A. Gerrard (Ed.), Protein nanotechnology: protocols, instrumentation and applications, 2nd ed., Humana Press, 2013, pp. 119–129 996.

[44] R. Mishra, K. Sorgjerd, S. Nystrom, A. Nordigarden, Y.-C. Yu, P. Hammarstrom, Lysozyme amyloidogenesis is accelerated by specific nicking and fragmentation but decelerated by intact protein binding and conversion, J Mol Biol 366 (2007) 1029–1044.

[45] S.G. Bolder, H. Hendrickx, L.M.C. Sagis, E. van der Linden, Fibril assemblies in aqueous whey protein mixtures, J Agric Food Chem 54 (2006) 4229–4234.

[46] S.G. Bolder, H. Hendrickx, L.M.C. Sagis, E. Van der Linden, Ca^{2+}-induced cold-set gelation of whey protein isolate fibrils, Appl Rheol 16 (2006) 258–264.

[47] S.M. Loveday, J. Su, M.A. Rao, S.G. Anema, H. Singh, Effect of calcium on the morphology and functionality of whey protein nanofibrils, Biomacromolecules 12 (2011) 3780–3788.

[48] J.T. de Faria, V.P. Rodrigues Minim, L.A. Minim, Evaluating the effect of protein composition on gelation and viscoelastic characteristics of acid-induced whey protein gels, Food Hydrocoll 32 (2013) 64–71.

[49] J. Healy, K. Wong, E.B. Sawyer, C. Roux, L. Domigan, S.L. Gras, et al. Polymorphism and higher order structures of protein nanofibers from crude mixtures of fish lens crystallins: toward useful materials, Biopolymers 97 (2012) 595–606.

[50] C. Valery, F. Artzner, M. Paternostre, Peptide nanotubes: molecular organisations, self-assembly mechanisms and applications, Soft Matter 7 (2011) 9583–9594.

[51] C. Valery, R. Pandey, J.A. Gerrard, Protein beta-interfaces as a generic source of native peptide tectons, Chem Commun 49 (2013) 2825–2827.

[52] J. Castillo, L. Sasso, W.E. Svendsen, Self-assembled peptide nanostructures: advances and applications in nanobiotechnology, Pan Stanford, (2012).

[53] M. Reches, E. Gazit, Casting metal nanowires within discrete self-assembled peptide nanotubes, Science 300 (2003) 625–627.

[54] L. Adler-Abramovich, D. Aronov, P. Beker, M. Yevnin, S. Stempler, L. Buzhansky, et al. Self-assembled arrays of peptide nanotubes by vapour deposition, Nat Nanotechnol 4 (2009) 849–854.

[55] J. Ryu, C.B. Park, Solid-phase growth of nanostructures from amorphous peptide thin film: effect of water activity and temperature, ChemMat 20 (2008) 4284–4290.

[56] J. Ryu, C.B. Park, High-temperature self-assembly of peptides into vertically well-aligned nanowires by aniline vapor, Adv Mat 20 (2008) 3754–3758.

[57] J. Ryu, J.S. Lee, C.B. Park, Self-assembly of diphenylalanine peptides into nanowires for nanobiotechnology applications, Abstr Pap Am Chem Soc (2010) 239.

[58] L. Adler-Abramovich, N. Kol, I. Yanai, D. Barlam, R.Z. Shneck, E. Gazit, et al. Self-assembled organic nanostructures with metallic-like stiffness, Angew Chem -Int Edit 49 (2010) 9939–9942.

[59] I. Azuri, L. Adler-Abramovich, E. Gazit, O. Hod, L. Kronik, Why are diphenylalanine-based peptide nanostructures so rigid? Insights from first principles calculations, J Am Chem Soc 136 (2014) 963–969.

[60] M. Yemini, M. Reches, E. Gazit, J. Rishpon, Peptide nanotube-modified electrodes for enzyme-biosensor applications, Anal Chem 77 (2005) 5155–5159.

[61] M. Yemini, M. Reches, J. Rishpon, E. Gazit, Novel electrochemical biosensing platform using self-assembled peptide nanotubes, Nano Lett 5 (2005) 183–186.

[62] L. Adler-Abramovich, M. Badihi-Mossberg, E. Gazit, J. Rishpon, Characterization of peptide-nanostructure-modified electrodes and their application for ultrasensitive environmental monitoring, Small 6 (2010) 825–831.

[63] T.C. Cipriano, P.M. Takahashi, D. de Lima, V.X. Oliveira, J.A. Souza, H. Martinho, et al. Spatial organization of peptide nanotubes for electrochemical devices, J Mater Sci 45 (2010) 5101–5108.

[64] L. Soleymani, Z.C. Fang, X.P. Sun, H. Yang, B.J. Taft, E.H. Sargent, et al. Nanostructuring of patterned microelectrodes to enhance the sensitivity of electrochemical nucleic acids detection, Angew Chem -Int Edit 48 (2009) 8457–8460.

[65] H. Yang, S.Y. Fung, M. Pritzker, P. Chen, Ionic-complementary peptide matrix for enzyme immobilization and biomolecular sensing, Langmuir 25 (2009) 7773–7777.

[66] E.C. Cho, J.-W. Choi, M. Lee, K.-K. Koo, Fabrication of an electrochemical immunosensor with self-assembled peptide nanotubes, Colloid Surf A 313 (2008) 95–99.

[67] R. de la Rica, E. Mendoza, L.M. Lechuga, H. Matsui, Label-free pathogen detection with sensor chips assembled from peptide nanotubes, Angew Chem -Int Edit 47 (2008) 9752–9755.

[68] B.-W. Park, R. Zheng, K.-A. Ko, B.D. Cameron, D.-Y. Yoon, D.-S. Kim, A novel glucose biosensor using bi-enzyme incorporated with peptide nanotubes, Biosens Bioelectron 38 (2012) 295–301.

[69] L. Sasso, I. Vedarethinam, J. Emneus, W.E. Svendsen, J. Castillo-Leon, Self-assembled diphenylalanine nanowires for cellular studies and sensor applications, J Nanosci Nanotechnol 12 (2012) 3077–3083.

[70] W.E. Svendsen, J. Castillo-Leon, J.M. Lange, L. Sasso, M.H. Olsen, M. Abaddi, et al. Micro and nano-platforms for biological cell analysis, Sens Actuat A Phys 172 (2011) 54–60.

[71] W.E. Svendsen, J. Castillo-Leon, J.M. Lange, L. Sasso, M.H. Olsen, M. Abaddi, et al. Micro and nano-platforms for biological cell analysis, Eurosensors XXIV Conf 5 (2010) 33–36.

[72] M.B. Taskin, L. Sasso, M. Dimaki, W.E. Svendsen, J. Castillo-Leon, Combined cell culture-biosensing platform using vertically aligned patterned peptide nanofibers for cellular studies, ACS Appl Mater Interf 5 (2013) 3323–3328.

[73] I.D. Matos, W.A. Alves, Electrochemical determination of dopamine based on self-assembled peptide nanostructure, ACS Appl Mater Interf 3 (2011) 4437–4443.

[74] S. Kim, S.Y. Bae, B.Y. Lee, T.D. Kim, Coaggregation of amyloid fibrils for the preparation of stable and immobilized enzymes, Anal Biochem 421 (2012) 776–778.

[75] T. Scheibel, Protein fibers as performance proteins: new technologies and applications, Curr Opin Biotechnol 16 (2005) 427–433.

[76] W.F. Xue, A.L. Hellewell, W.S. Gosal, S.W. Homans, E.W. Hewitt, S.E. Radford, Fibril fragmentation enhances amyloid cytotoxicity, J Biol Chem 284 (2009) 34272–34282.

[77] A. Heredia, I. Bdikin, S. Kopyl, E. Mishina, S. Semin, A. Sigov, et al. Temperature-driven phase transformation in self-assembled diphenylalanine peptide nanotubes, J Phys D Appl Phys (2010) 43.

[78] R. Pauliukaite, M.E. Ghica, O. Fatibello, C.M.A. Brett, Comparative study of different cross-linking agents for the immobilization of functionalized carbon nanotubes within a chitosan film supported on a graphite-epoxy composite electrode, Anal Chem 81 (2009) 5364–5372.

[79] R.A. Sheldon, Enzyme immobilization: the quest for optimum performance, Adv Synth Catal 349 (2007) 1289–1307.

[80] J.I. Yeh, A. Lazareck, J.H. Kim, J. Xu, S. Du, Peptide nanowires for coordination and signal transduction of peroxidase biosensors to carbon nanotube electrode arrays, Biosens Bioelectr 23 (2007) 568–5674.

[81] S. Bolisetty, J. Adamcik, J. Heier, R. Mezzenga, Amyloid directed synthesis of titanium dioxide nanowires and their applications in hybrid photovoltaic devices, Adv Funct Mat 22 (2012) 3424–3428.

[82] C. Muller, R. Jansson, A. Elfwing, G. Askarieh, R. Karlsson, M. Hamedi, et al. Functionalisation of recombinant spider silk with conjugated polyelectrolytes, J Mat Chem 21 (2011) 2909–2915.

[83] M. Hamedi, A. Herland, R.H. Karlsson, O. Inganas, Electrochemical devices made from conducting nanowire networks self-assembled from amyloid fibrils and alkoxysulfonate PEDOT, Nano Lett 8 (2008) 1736–1740.

[84] M. Moehlenbrock, S. Minteer, Introduction to the field of enzyme immobilization and stabilization, in: S.D. Minteer (Ed.), Enzyme stabilization and immobilization, Humana Press, 2011, pp. 1–7 679.

[85] J. Castillo-León, W.E. Svendsen, M. Dimaki, Micro and nano techniques for the handling of biological samples, Taylor & Francis, (2011).

[86] P. Mesquida, E.M. Blanco, R.A. McKendry, Patterning amyloid peptide fibrils by AFM charge writing, Langmuir 22 (2006) 9089–9091.

[87] N. Nuraje, I.A. Banerjee, R.I. MacCuspie, L.T. Yu, H. Matsui, Biological bottom-up assembly of antibody nanotubes on patterned antigen arrays, J Am Chem Soc 126 (2004) 8088–8089.

[88] H. Yang, S.Y. Fung, M. Pritzker, P. Chen, Mechanical-force-induced nucleation and growth of peptide nanofibers at liquid/solid interfaces, Angew Chem -Int Edit 47 (2008) 4397–4400.

[89] J. Shklovsky, P. Beker, N. Amdursky, E. Gazit, G. Rosenman, Bioinspired peptide nanotubes: deposition technology and physical properties, Mater Sci Eng B Adv Funct Solid State Mater 169 (2010) 62–66.

[90] S.J. Tans, A.R.M. Verschueren, C. Dekker, Room-temperature transistor based on a single carbon nanotube, Nature 393 (1998) 49–52.

[91] W.E. Svendsen, M. Jørgensen, L. Andresen, K.B. Andersen, M.B.B.S. Larsen, S. Skov, et al. Silicon nanowire as virus sensor in a total analysis system, Proc Eng 25 (2011) 288–291.

[92] L.L. del Mercato, P.P. Pompa, G. Maruccio, A. Della Torre, S. Sabella, et al. Charge transport and intrinsic fluorescence in amyloid-like fibrils, Proc Natl Acad Sci USA 104 (2007) 18019–18024.

[93] M. Amit, G. Cheng, I.W. Hamley, N. Ashkenasy, Conductance of amyloid beta based peptide filaments: structure-function relations, Soft Matter 8 (2012) 8690–8696.

[94] J. Huang, H.L. Zhu, Y.C. Chen, C. Preston, K. Rohrbach, J. Cumings, et al. Highly transparent and flexible nanopaper transistors, ACS Nano 7 (2013) 2106–2113.

CHAPTER 2

Fabrication of Nanostructures Using Self-Assembled Peptides as Templates: The Diphenylalanine Case

Jaime Castillo-León

DTU Nanotech, Kgs. Lyngby, Technical University of Denmark, Lund, Sweden

2.1 INTRODUCTION

One-dimensional nanostructures such as silicon (Si) nanowires, carbon nanotubes (NTs), or III–V nanowires have been extensively used in numerous applications including in fields such as biosensing, electronics, communication, as well as for reinforced materials thanks to their electrical, photonic, optical, and mechanical properties. However, despite all the advantages of 1D nanostructures, their fabrication involves top-down and bottom-up approaches in which the use of specialized equipment, high energy, and in some cases clean-room facilities increases both the fabrication time and costs.

To overcome these challenges, new biomaterials and techniques are being explored. One alternative that has been looked into during the last two decades is the use of self-assembled biological entities as templates for the fabrication of novel nanostructures. An example of biological self-assembled materials is the short aromatic self-assembled peptide, diphenylalanine. This peptide is able to self-organize into various nanostructures (tubes, fibers, or solid particles) with properties that change depending on the formed arrangement. These nanostructures are formed under very mild conditions (aqueous medium, low temperatures, and outside clean-room facilities) in a couple of minutes.

This chapter presents the reasons for using this short aromatic dipeptide for the fabrication of nanostructures along with the methods to create these new nanostructures from the self-assembled peptide. Particularly, the fabrication of Ag nanowires, polymer nanochannels, Si, and conducting polymer nanowires using diphenylalanine as a template are explained in detail.

Micro and Nanofabrication Using Self-Assembled Biological Nanostructures. DOI: 10.1016/B978-0-323-29642-7.00002-3

2.2 DIPHENYLALANINE PEPTIDE

2.2.1 Synthesis of Nanostructures Using Diphenylalanine

As mentioned in Section 2.1, the short aromatic dipeptide diphenyl-alanine (FF) can be used for the synthesis of nanostructures such as NTs, nanofibers, or solid nanoparticles (NPs) under very mild conditions such as an aqueous environment, low temperatures, and without the need for specialized equipment [1]. In order to obtain any of the mentioned nanostructures, the synthesis parameters need to be adjusted as shown in Scheme 2.1; the first step of the synthesis is common to the formation of all three nanostructures and involves the dilution of the peptide powder in an alcohol, 1,1,1,3,3,3-hexafluoroisopropanol (HFP), at concentrations between 100 and 200 mg/mL. This alcohol is one of the few solvents that are able to completely and rapidly dissolve this dipeptide, giving rise to a transparent solution of the dissolved compound. For some applications, its use could bring additional challenges since it is known to attack materials commonly utilized in micro- and nanofabrication (e.g., polymers). Several alternative syntheses, in the absence of HFP, have been published, but these require higher energies and longer fabrication times [2–4].

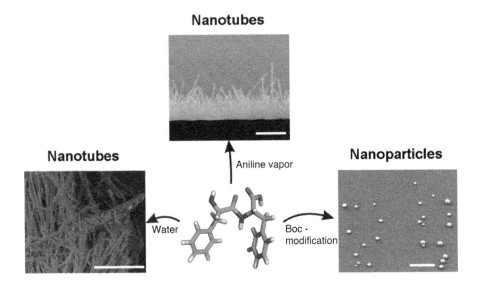

Nanotubes

Nanotubes   **Nanoparticles**

Aniline vapor

Water   Boc - modification

Diphenylalanine

Scheme 2.1. Self-assembled nanostructures synthesized using diphenylalanine.

Once the FF stock solution is obtained, the subsequent step is to blend with water or a mix of water/ethanol, or carry out an exposure to aniline vapor in order to obtain the desired nanostructures. In the case of forming NTs and NPs, the self-assembly of these nanostructures takes a couple of seconds, and hundreds of structures are rapidly obtained. This fast formation leads to nanostructures of varying lengths and diameters, which brings the extra challenge of controlling these dimensions when required. An alternative to solve this problem is to use templates or on-chip fabrication, which offers a more controlled mixing of the required components, resulting in the formation of nanostructures with less dispersed sizes [5].

2.2.2 Stability of Diphenylalanine Nanostructures

Despite the extraordinary mechanical properties reported for FF NTs and NPs (Young's modulus values of 19 and 275 GPa, respectively, and a high bending stiffness for the NTs) [6,7], these nanostructures dissolve very quickly when submerged in organic solvents such as water, phosphate buffer, or ethanol [8], cf. Figure 2.1. This behavior, which could be a limitation for some applications, is the key for the successful use of these peptide structures as etching masks in the fabrication of semiconducting nanowires. On the other hand, it was demonstrated that despite being formed using the same dipeptide, nanofibers formed using FF were not dissolved in water, as was the case for NTs and NPs. Until now, researchers have yet to come up with a definitive explanation to this behavior, but several studies discussing this difference in stability are being reported [8–10].

Apart from the differences in stability in liquid environments, FF NTs and nanofibers also behave differently when exposed to the bombardment of ions in a technique known as reactive-ion etching (RIE) [11]. This method combines physical and chemical effects to remove material from a surface [12]. Figure 2.2 shows the difference in behavior between FF NTs and their nanofiber counterparts when exposed to RIE; while the NTs could tolerate up to 20 s of ion bombardment, the nanofibers were etched very quickly.

These differences in behavior between the nanostructures formed using FF offer several alternatives for their use in micro- and nanofabrication, highlighting the advantages of employing this biological material as an alternative to traditional fabrication methods for the synthesis of micro- and nanostructures.

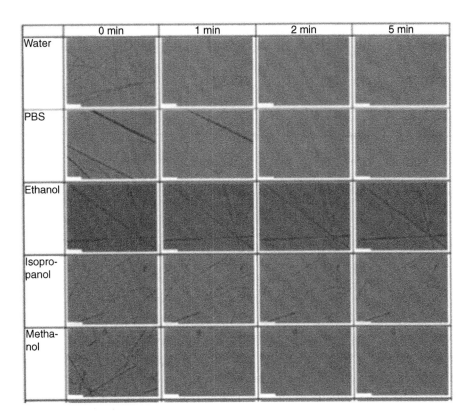

Fig. 2.1. Micrographs of peptide nanotubes being dissolved after submersion in distilled water, phosphate buffer (pH 7.4), ethanol, isopropanol, and methanol. All scale bars in the images correspond to 20 μm. Reproduced from Ref. [8] with permission from The Royal Society of Chemistry.

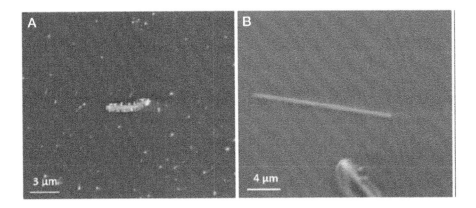

Fig. 2.2. Comparison of the stability of self-assembled diphenylalanine nanofibers (A) and nanotubes (B) after 60 s under a reactive-ion etching process [11]. Reproduced with permission from Springer Science and Business Media.

2.2.3 Manipulation of Diphenylalanine Peptide Nanostructures

In some cases, the self-assembled nanostructure used as template needs to be positioned in a specific location before the etching process can take place. For this, it is necessary to use manipulation techniques that render it possible to precisely move and position the biological nanostructure to a desired location. The handling of biological nanostructures requires the use of techniques that do not affect their structure [13]. As mentioned in Section 2.2.2, FF NTs dissolve rapidly when submerged in common organic solvents, which can be a limitation when using manipulation techniques such as microfluidics, dielectrophoresis, or acoustophoresis, which require the use of liquids for handling the nanostructure. One solution could be the use of solvents that do not interact with the structure of the self-assembled entity. Apart from the three abovementioned manipulation techniques, other contact and noncontact handling techniques are available and are summarized in Table 2.1.

2.3 FABRICATION OF NEW NANOSTRUCTURES USING DIPHENYLALANINE NANOSTRUCTURES AS A TEMPLATE

2.3.1 Fabrication of Metallic Nanowires and Coaxial Cables

Reches and Gazit [23] made use of the internal cavity present in FF NTs and developed a very simple yet clever method for casting Ag nanowires. This casting of metallic nanowires starts with the preparation of an FF stock solution in HFP at a concentration of 100 mg/mL as explained in Section 2.2.1. The peptide stock solution is then diluted with water

Table 2.1 Manipulation/Immobilization Techniques for the Handling of Self-assembled Peptides		
Techniques	**Details**	**References**
Dielectrophoresis	Noncontact method, sample in liquid	[14,15]
Magnetic alignment	Noncontact method, sample needs to be magnetic or to be modified with magnetic NPs	[16,17]
Inkjet printing	Noncontact method, sample in solution	[18]
Spin casting	Noncontact method, sample in liquid	[19]
Atomic force microscopy (AFM)	Contact method	[20,21]
Thiol-based immobilization	Sample needs to be functionalized with thiol groups	[22]

to a final concentration of 2 mg/mL. The formation of a white precipitate can be immediately observed with the naked eye, indicating the self-assembly of FF NTs. For the casting of Ag nanowires, an aliquot (90 μL) of FF NTs is added to 10 μL of a boiling solution of AgNO₃ (10 mM). Following this step, citric acid, a reducing agent, is added to obtain a final concentration of 0.038%, which causes the reduction of silver (Ag) inside the FF NT thus creating a metallic nanowire with a diameter similar to that of the FF internal cavity. The next step involves the elimination of the peptide layer to obtain the Ag nanowire. This is done by incubating the Ag-filled FF NTs with proteinase K at a final concentration of 100 μg/mL for 1 h at 37°C.

A variation of the method invented by Reches and Gazit was developed by the same group in order to obtain coaxial metal nanocables [24]. In this case, the FF stock solution was mixed with linker peptides in order to introduce linker groups on the external surface of the FF NTs to later cover the NTs with a layer of Au NPs. In this approach, the peptide NT structure was kept after the reduction of Ag within the FF NT cavity. Cys-Gly-Ser-Phe-Phe and Phe-Phe-Cys peptides purchased from Bachem (Switzerland) were used as linker peptides. The lyophilized form of these peptides was dissolved in dithiothreitol (0.25 M) to a final concentration of 25 mg/mL. The peptide linker and the FF stock solution were mixed at 1:10 ratio. The reduction of Ag was performed as described above. For the binding of gold (Au) NPs, monomaleimido nanogold with a diameter of 1.4 nm (Nanoprobes, USA) was obtained by cross-linking the nanogold to linker peptides at 4°C for 18 h. The formed nanostructures were metal–insulator–metal, trilayer coaxial nanocables, cf. Figure 2.3.

2.3.2 Fabrication of Polymer Nanochannels

The size of channels fabricated by traditional micro- and nanofabrication techniques may be limited depending on the fabrication technique used, and an alternative to obtain channels with diameters in the nano-range is to use biological nanostructures with internal cavities below 100 nm. A nano-fluidic channel can be created by using an FF NT as the template, taking advantage of its internal cavity for liquid to pass through. To do this, FF NTs were fabricated as described in Section 2.2.1, and placed on a desired surface, for example, connecting two previously fabricated containers. Once the NT was in the desired location, the FF

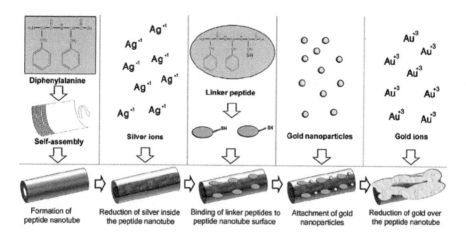

| Diphenylalanine | Silver ions | Linker peptide | Gold nanoparticles | Gold ions |

| Formation of peptide nanotube | Reduction of silver inside the peptide nanotube | Binding of linker peptides to peptide nanotube surface | Attachment of gold nanoparticles | Reduction of gold over the peptide nanotube |

Fig. 2.3. Fabrication of a coaxial nanocable using an FF peptide NT as a template for the reduction of Ag ions in the internal cavity of the NT and the reduction of Au over its external surface. The final nanostructure was a metal–insulator–metal trilayer coaxial cable. Reproduced with permission from Ref. [24]. Copyright (2006) American Chemical Society.

nanostructure was covered by a polymer such as poly-dimethyl-siloxane (PDMS), which was subsequently cured creating the microfluidic nano-channel. The diameter of the nanochannel was defined by the diameter of the FF NT, which could vary between 80 and 300 nm. Sopher et al. used this approach to present the fabrication of a nanochannel for the transport of liquids between two microcontainers [25]. Again, one should take into account that, depending on the liquid transport through the NT, its structure could be affected if the liquid is an organic solvent such as water. Nevertheless, the polymer nanochannel structure will be retained even if the FF NT is dissolved.

2.3.3 Fabrication of Silicon and Conducting Polymer Nanowires

A more simple fabrication of Si and conducting polymer nanowires was developed by using FF NTs as a dry etching mask in a deep RIE fabrication process [11,19,26]. The key to the success of this method is the capacity of the FF NTs to withstand the RIE process and their solubility in water. This renders it possible to avoid the use of organic solvents such as acetone to remove the etching mask, thus leaving behind the formed nanostructure. This method is faster, cheaper, and more environment-friendly than traditional fabrication techniques to obtain Si and conducting polymers nanowires.

The fabrication of Si nanowires using FF NTs as a dry etching mask is stated, as in the previously explained fabrication methods, with the synthesis of FF NTs as explained in Section 2.2.1, after which the NTs are placed on top of a Si wafer. This positioning of the FF NT can be done in a more precise way by using some of the methods mentioned in Table 2.1. Once the solvent is evaporated and the dry FF NTs are placed in the desired location, the Si wafer covered with FF NTs is placed inside an STS C010 multiplex cluster system for the reactive etching process. SF_6 and O_2 with respective flow rates of 32 and 8 sccm are used. The pressure inside the vacuum chamber is adjusted to 80 mTorr and the RF power to 30 W. The wafer is etched for 5 min under these conditions. After etching, the processed wafer is placed in a distilled water bath to remove the FF NT acting as an etching mask, leaving the formed Si nanostructure with the same length and diameter as the FF NT. Thus, Si nanowires with diameters between 80 and 300 nm can be obtained in less than 10 min outside clean-room facilities. The whole process is summarized in Figure 2.4.

For the fabrication of conducting polymer nanowires, a Si oxide wafer can be spin coated with a layer of p-toluenesulfonate-doped poly(3,4-ethylenedioxythiophene) (PEDOT:TsO) with a spin rate of 4000 rpm for 60 s. The coated wafer is heated to 70°C for 10 min in order

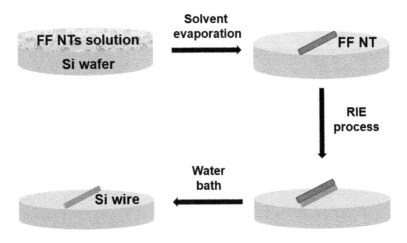

Fig. 2.4. Process steps for the fabrication of Si nanostructures using an FF peptide NT as an etching mask under an RIE etching process. Once the RIE process is finished, the FF NT acting as a dry etch mask is removed by dipping the wafer in a water bath where the NT is dissolved, leaving the Si nanostructure.

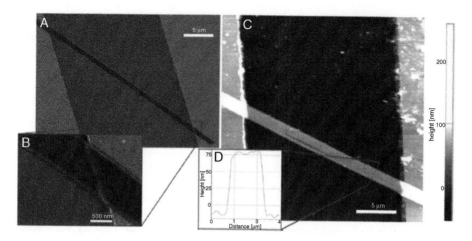

Fig. 2.5. Micrographs from scanning electron microscopy (A and B) and AFM (C) of the fabricated PEDOT:TsO nanowire structures fabricated using an FF NT as the template under a RIE process. (A) A SEM micrograph of the whole PEDOT:TsO nanowire spanning the gap between two Au electrodes. (B) A zoom of the contact area between the PEDOT:TsO nanowire and the Au contacts. (C) An AFM image of the junction between the PEDOT:TsO nanowire and the Au contact pad. (D) A line scan covering the nanowire showing the height (75 nm) of the fabricated nanostructure. Reprinted from Ref. [26] with permission from Elsevier.

to evaporate the inhibitor pyradine and start the polymerization process. Subsequently, the FF NTs are placed on top of the PEDOT:TsO-coated wafer at a desired location. Following this, the modified wafer is placed in a STS C010 multiplex cluster system with a pressure of 300 mTorr and a power of 100 W. For the patterning of the PEDOT:TsO, oxygen-based plasma is used (98 sccm O_2 and 20 sccm N_2) for 15 s, after which the wafer is placed in a water bath to remove the dry etching FF NT mask. Using this method, PEDOT nanowires aimed to be used as temperature sensors were fabricated [26] (Figure 2.5).

2.4 CONCLUSIONS

Self-assembled biological nanostructures are materials that offer numerous advantages for the rapid fabrication of new micro- and nanostructures in a more rapid, low-cost, and environment-friendly fashion. The FF peptide is a clear example of this; by using the nanostructures based on this short aromatic dipeptide, 1D metallic and semiconducting polymer nanowires as well as polymer nanochannels have been fabricated. The presented methods are the result of a multidisciplinary effort involving researchers from different disciplines such as biology, chemistry, physics, and engineering in order to find new materials and methods to overcome the challenges that exist with traditional fabrication techniques.

REFERENCES

[1] M. Reches, E. Gazit, Formation of closed-cage nanostructures by self-assembly of aromatic dipeptides, Nano Lett 4 (2004) 581–585.

[2] L. Adler-Abramovich, D. Aronov, P. Beker, M. Yevnin, S. Stempler, L. Buzhansky, et al. Self-assembled arrays of peptide nanotubes by vapour deposition, Nat Nanotechnol 4 (2009) 849–854.

[3] J. Kim, T.H. Han, Y. Kim, J.S. Park, J. Choi, D.G. Churchill, et al. Role of water in directing diphenylalanine assembly into nanotubes and nanowires, Adv Mater 22 (2010) 583–587.

[4] J. Ryu, C.B. Park, High-temperature self-assembly of peptides into vertically well-aligned nanowires by aniline vapor, Adv Mater 20 (2008) 3754–3758.

[5] J. Castillo-León, R. Rodriguez-Trujillo, S. Gauthier, A.C.Ø. Jensen, W.E. Svendsen, Micro-"factory" for self-assembled peptide nanostructures, Microelectron Eng 88 (2011) 1685–1688.

[6] L. Adler-Abramovich, N. Kol, I. Yanai, D. Barlam, R.Z. Shneck, E. Gazit, et al. Self-assembled organic nanostructures with metallic-like stiffness, Angew Chem Int Ed 49 (2010) 9939–9942.

[7] N. Kol, L. Adler-Abramovich, D. Barlam, R.Z. Shneck, E. Gazit, I. Rousso, Self-assembled peptide nanotubes are uniquely rigid bioinspired supramolecular structures, Nano Lett 5 (2005) 1343–1346.

[8] K.B. Andersen, J. Castillo-Leon, M. Hedstrom, W.E. Svendsen, Stability of diphenylalanine peptide nanotubes in solution, Nanoscale 3 (2011) 994–998.

[9] T.O. Mason, D.Y. Chirgadze, A. Levin, L. Adler-Abramovich, E. Gazit, T.P.J. Knowles, et al. Expanding the solvent chemical space for self-assembly of dipeptide nanostructures, ACS Nano 8 (2014) 1243–1253.

[10] J. Ryu, C.B. Park, High stability of self-assembled peptide nanowires against thermal, chemical, and proteolytic attacks, Biotechnol Bioeng 105 (2010) 221–230.

[11] M.B. Larsen, K.B. Andersen, W.E. Svendsen, J. Castillo-León, Self-assembled peptide nanotubes as an etching material for the rapid fabrication of silicon wires, BioNanoSci 1 (2011) 31–37.

[12] F. Laermer, S. Franssila, L. Sainiemi, K. Kolari, Deep reactive ion etching, in: V. Lindroos, M. Tilli, A. Lehto, T. Mottoka (Eds.), Handbook of silicon based MEMS materials and technologies, Elsevier, Oxford, 2010, pp. 349.

[13] J. Castillo, M. Dimaki, W.E. Svendsen, Manipulation of biological samples using micro and nano techniques, Integr Biol 1 (2009) 30–42.

[14] J. Castillo, S. Tanzi, M. Dimaki, W. Svendsen, Manipulation of self-assembly amyloid peptide nanotubes by dielectrophoresis, Electrophoresis 29 (2008) 5026–5032.

[15] R. de la Rica, E. Mendoza, L.M. Lechuga, H. Matsui, Label-free pathogen detection with sensor chips assembled from peptide nanotubes, Angew Chem Int Ed 47 (2008) 9752–9755.

[16] R.J.A. Hill, V.L. Sedman, S. Allen, P.M. Williams, M. Paoli, L. Adler-Abramovich, et al. Alignment of aromatic peptide tubes in strong magnetic fields, Adv Mater 19 (2007) 4474–4479.

[17] M. Reches, E. Gazit, Controlled patterning of aligned self-assembled peptide nanotubes, Nat Nanotechnol 1 (2006) 195–200.

[18] L. Adler-Abramovich, E. Gazit, Controlled patterning of peptide nanotubes and nanospheres using inkjet printing technology, J Pept Sci 14 (2008) 217–223.

[19] K.B. Andersen, J. Castillo-León, T. Bakmand, W.E. Svendsen, Alignment and use of self-assembled peptide nanotubes as dry-etching mask, Jap J Appl Phys 51 (2012) 106FF131–1106FF.

[20] P. Mesquida, E.M. Blanco, R.A. McKendry, Patterning amyloid peptide fibrils by AFM charge writing, Langmuir 22 (2006) 9089–9091.

[21] V.L. Sedman, L. Adler-Abramovich, S. Allen, E. Gazit, S.J.B. Tendler, Direct observation of the release of phenylalanine from diphenylalanine nanotubes, J Am Chem Soc 128 (2006) 6903–6908.

[22] B. Viguier, K. Zor, E. Kasotakis, A. Mitraki, C.H. Clausen, W.E. Svendsen, et al. Development of an electrochemical metal-ion biosensor using self-assembled peptide nanofibrils, ACS Appl Mater Interfaces 3 (2011) 1594–1600.

[23] M. Reches, E. Gazit, Casting metal nanowires within discrete self-assembled peptide nanotubes, Science 300 (2003) 625–627.

[24] O. Carny, D.E. Shalev, E. Gazit, Fabrication of coaxial metal nanocables using a self-assembled peptide nanotube scaffold, Nano Lett 6 (2006) 1594–1597.

[25] N.B. Sopher, Z.R. Abrams, M. Reches, E. Gazit, Y. Hanein, Integrating peptide nanotube in micro-fabrication processes, J Micromech Microeng 17 (2007) 2360–2365.

[26] K.B. Andersen, N.O. Christiansen, J. Castillo-León, N. Rozlosnik, W.E. Svendsen, Fabrication and characterization of PEDOT nanowires based on self-assembled peptide nanotube lithography, Org Electron 14 (2013) 1370–1375.

Self-Assembled Peptide Nanostructures for the Fabrication of Cell Scaffolds

Rui Li[1], Alexandra Rodriguez[2], David R. Nisbet[2], Colin J. Barrow[1], Richard J. Williams[1]

[1]Centre for Chemistry and Biotechnology, Deakin University, Victoria, Australia

[2]Research School of Engineering, The Australian National University, Canberra, Acton, Australia

3.1 INTRODUCTION

Regenerative medicine and tissue engineering have emerged as important avenues for the treatment of many human diseases with the purpose of replacing, repairing, or enhancing the biological function of damaged tissues and/or organs [1]. To achieve such complex tissue repair, it is necessary to generate a three-dimensional (3D) microenvironment that possesses the appropriate biochemical and biomechanical properties that closely mimic the extracellular niche in order to provide structural and bioactive support to exogenous and endogenous cells, ultimately regenerating the damaged tissue [2]. The design of materials with well-defined properties is of interest, especially in forming nanoscale assemblies that can provide a scaffolding support for a range of applications including regenerative medicine [3].

The native microenvironment is composed of the extracellular matrix (ECM), an aqueous milieu with a self-assembled interlocking 3D structure, composed of a dynamic and tissue-specific range of signaling molecules including proteins (e.g., fibronectin and laminin), collagen, glycosaminoglycans, and proteoglycans (PGs), enriched with soluble factors [4]. This scaffold-like support provides the bulk, shape, and strength of many tissues *in vivo* [5]. While soluble morphogens play an important role in controlling cellular processes, it has been shown that ECM peptide-binding domains are specifically recognized by integrin cell surface receptors and have profound effects on cell migration, survival, proliferation, and differentiation [6]. Additionally, cells can "sense" the mechanical properties of the ECM, further directing cell behavior [7].

Micro and Nanofabrication Using Self-Assembled Biological Nanostructures. DOI: 10.1016/B978-0-323-29642-7.00003-5

For successful tissue regeneration, scaffolds should be able to emulate the ECM both spatially and temporally in order to control cell survival, proliferation, migration, and differentiation of existing and introduced cells with the ultimate goal of promoting tissue repair and regeneration. One effective approach to design such materials is by molecular self-assembly, a process which is ubiquitous in nature [8], thereby providing a template for the design of complex nanostructures based on simple interactions as seen in nature, leading to significant advances in molecular engineering, advanced biomaterials, and nanomaterials [9,10].

Among such biomaterials, self-assembled peptides (SAPs) are a promising candidate for applications in regenerative medicine [11]. Peptide hydrogels are hydrophilic, highly hydrated, 3D, polymeric networks polymers that mechanically resemble a number of biological tissues [12], and provide many specific features through the presence of biochemical and biomechanical signals in a matrix similar to the extracellular microenvironment [13]. The use of SAPs as building blocks to engineer biofunctional nanofibrous scaffolds holds many advantages including their inherent biocompatibility, simplicity, low cost, and ease of synthesis using standard peptide synthesis approaches such as solid-phase peptide synthesis (SPPS) [14]. The information for self-assembly is coded in the amino acid sequence, enabling it to be tuned and form a diverse range of structural characteristics at the nanoscale level, allowing dynamic control of the target material organization and properties [15]. Many types of architectures such as vesicles, micelles, monolayers, bilayers, fibers, ribbons, and tapes have been engineered by SAPs [16]. Furthermore, SAPs have a long shelf-life under room temperature, which is necessary for translation to the clinic [11]. Here we review the various existing SAP systems and their development for use in tissue engineering.

3.2 CLASSES OF SELF-ASSEMBLED PEPTIDE SCAFFOLDS

Proteins are large macromolecules consisting of a genetically encoded sequence of amino acids, linked together into a polypeptide chain. There are 20 natural amino acids, used as a toolbox in nature, each providing a different chemistry via changes in side chain functional groups.

Structurally, these amino acids provide a range of noncovalent inter-molecular forces including hydrogen bonding, hydrophobic interactions, ionic bonds, electrostatic interactions, π-stacking, van der Waals inter-actions and the formation of covalent disulfide bridges [17]. Although individually these forces are quite weak, collectively they give rise to stable secondary and tertiary protein structures.

Researchers have taken these sequences and used the structures they form to drive the formation of materials. Generally speaking, there are two categories of SAP systems, natural and nonnatural. Natural SAP systems form α-helices, coiled coils, β-sheets, and β-hairpins through the utilization of basic conformational units of naturally existing pro-teins. In contrast, nonnatural SAP systems involve the covalent linking of a sequence of amino acids to another (semi)synthetic molecule such as an alkyl chain or aromatic group to stabilize the formation of nano-fibers via noncovalent interactions [14].

3.3 NATURAL SELF-ASSEMBLING PEPTIDE SYSTEMS

3.3.1 α-Helices/Coiled Coils

The α-helical coiled-coil is a versatile repeating motif found in pro-teins [18]. It is a common secondary structure of proteins and usually takes the form of a right-handed coil or spiral conformation in which every C = O group (residue n) of the peptide backbone is hydrogen-bonded to the backbone amide hydrogen of the fourth residue further toward the C-terminus (residue $n + 4$). The primary helix–helix in-teractions can form higher-order structures and thickened fibers that have been found in a variety of natural coiled-coil proteins. Woolfson and colleagues engineered self-assembling fibers that formed α-helical coiled-coils [19–22]. These fibers were developed using two comple-mentary 28-residue peptides containing a heptad sequence repeat in the self-assembling fibers, "abcdefg." Isoleucine and leucine were located in the a and d positions, respectively [19]. The overall structure is amphipathic and oligomerizes through hydrophobic interactions. Gelation of this system can be controlled by inducing fiber formation through the controlled mixing of the complementary components, resulting in fibers that are a highly ordered precipitate formed from solution. For the formation of hydrogelating self-assembled fibers with >99% water content, the peptide sequences are engineered by altering the b, c, and f sites, to alanine in order to enhance weak hydrophobic interactions between fibrils or glutamine to promote hydrogen bonding. Through this mechanism, smaller, more flexible bundles of thinner fibers that underpin the spanning networks were formed [23].

3.3.2 β-Sheets

β-Sheets are composed of β-strands that contain multiple peptide chains to provide extended backbone arrangements that allow hydrogen bonding between the backbone amides and carbonyls. A β-strand is a stretch of peptide chain that typically consists of 3–10 amino acids with the backbone in an almost fully extended conformation. In nature, β-sheets have two arrangements: in parallel, where amino acid chains are arranged from the N-terminal to C-terminal, or antiparallel patterns where the chains N and C terminal are in reverse direction (Figure 3.1.A). The resulting pattern produces strong interstrand stability through planar hydrogen bonds between carbonyls and amines [14]. This basic motif results in alternating hydrophobic and hydrophilic residues so that the sheet has two faces: hydrophobic and hydrophilic. This allows the β-sheets to come together and conceal the hydrophobic surface from surrounding water molecules.

The major challenge of engineering a self-assembling fibrous system is to control the assembly conditions in order to form consistent reproducible structures. A variety of hierarchical structures can be formed from β-sheet peptides, such as tapes, ribbons, fibrils, and fibers [24]. These different structures are determined by the peptide and salt concentrations with higher-level assemblies formed as the peptide concentration is increased. In 1993, Zhang et al. first found that the Zuotin protein from yeast adopted β-sheet structure by nature of self-complementary ionic interactions and led to the formation of nanofibers, naming it EAK16 (AEAEAKAKAEAEAKAK) for its amino acid composition [25]. Based on this design, similar peptides were engineered to form 3D nanofibrous self-assembling scaffolds such as RADA16 (acteyl-RARADADAARARADADA-CONH$_2$), consisting of an alternating hydrophobic and hydrophilic amino acid sequence that forms a stable β-sheet structure. More recently, the octapeptide FEFEFKFK was engineered to self-assembled based on the same ionic-complementary interactions [26]. The alternation of nonpolar hydrophobic and polar hydrophilic residues formed transparent and self-supporting hydrogels by increasing the temperature to 90°C in order to dissolve the peptide and then cooling the solution to room temperature to form a hydrogel rich in β-sheet nanofibers. The critical gelation concentration was 5 mg/mL and the gel was formed within 1 min [26]. The disadvantage of such SAP systems, however, is the length of the peptide sequence (>12 amino acids), increasing complexity to the synthesis process resulting in increased cost and reduction in purity.

3.3.3 β-Hairpins

The β-hairpin structure consists of two short β-strands connected by a short loop of two to five amino acids. The β-strands are adjacent in primary structure, oriented in an antiparallel direction. Two 20-amino-acid-peptides MAX1 and MAX8 have been designed, both of which can undergo gelation initiated by Dulbecco's modified Eagle medium (DMEM) cell culture media [26,27] (Figure 3.1.B). Due to a change in overall peptide charge, the gelation speed of MAX8 under physiological conditions is faster than the original MAX1 sequence, which is important for desired, homogeneous, 3D cell encapsulation, and injectable delivery of cells *in vivo* [28].

3.4 SEMISYNTHETIC SELF-ASSEMBLING PEPTIDE SYSTEMS

3.4.1 Peptide Amphiphiles

A characteristic property of peptide amphiphiles (PAs) is that the subunit consists of a polymer linked to a peptide. This provides a hydrophobic aliphatic tail of a tuneable length and a hydrophilic peptide sequence connected to that tail through an amide bond. The hydrophobic tail is usually 12–16 carbon alkyl subunits long. Upon application of a trigger such as a change in pH or ionic strength, these PA molecules can self-assemble into various supramolecular-structures including nanotapes [24], ribbons [24,29], fibers [30], and twisted ribbons [31]. The formation of nanofibers arise when the alkyl group forms a hydrophilic core and the four amino acids adjacent to the alkyl group form stable β-sheet structures [32,33]. The remainder of the amino acids displayed on the periphery provides the potential

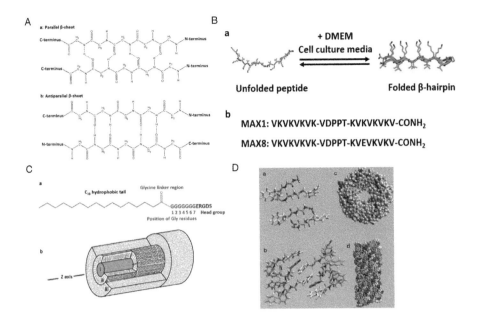

Fig. 3.1. Formation of structure using self-assembling peptide motifs. (A) The structure of (a) parallel and (b) antiparallel β-sheets. (B) (a) The self-assembly of MAX1 and MAX8 via a β-hairpin mechanism. (b) The peptide sequences of MAX1 and MAX8. Adapted with permission from Ref. [26]. Copyright (2005) Elsevier. *(C) (a) Schematic representation of a PA. (b) 3D representation of the regions of a PA nanofiber. Region (i) is the hydrophobic core composed of aliphatic tails. Region (ii) is the critical β-sheet hydrogen bonding part of the peptide and consists of four amino acids. Region (iii) is the peripheral peptide region which is not constrained to a particular hydrogen-bonding motif and forms the interface with the environment.* Adapted with permission from Ref. [34]. Copyright (2006) American Chemical Society. *(D) (a) The molecular structure of Fmoc-FF with the formation of antiparallel β-sheet pattern. (b) Fmoc groups undergo π-stacking to create a backbone. (c) Tubes are formed via the curve of the peptide sequence. (d) The assembly forms long stacks of these, resulting in the fibrillar formation. Dark gray (orange in the web version), fluorenyl groups; light gray (purple in the web version), phenyl groups.* Adapted with permission from Ref. [45]. Copyright (2008) WILEY-VCH Verlag GmbH & Co. KGaA, Weinheim.

for the presentation of bioactive peptide sequences at high density (Figure 3.1.C) that can act as biochemical cues for cell support and direction of cell fate [34]. However, these PAs are synthetically complex, the high cost for synthesis limiting their potential for large-scale production.

3.4.2 Lipid-Like Peptides

Recently, Hauser et al. have shown a new type of aliphatic peptide that can self-assemble into helical fibers in supramolecular structures. The specific sequence of the novel peptides contain aliphatic amino acids with decreasing hydrophobicity capped with a hydrophilic head group, such as acetyl-AIVAGD [35]. Importantly, the assembly process of these minimalist peptides change from random coil to α-helical intermediates and finally to cross-β spins. After this, the peptide pairs assembled into fibers and the condensation of fibers forms a hydrogel. The resulting 3D hydrogel scaffold was able to entrap up to 99.9% water and resemble collagen fibers in the ECM [35,36]. They were also found to be heat-resistant up to 90°C and have high and tunable mechanical strength.

3.4.3 Aromatically Capped Peptide Derivatives

Amino acids containing aromatic side chains have key roles in the self-assembly process by forming π–π stacks [37]. Much research has therefore been focused on Fmoc-protected peptides as the N-terminally protecting 9-fluorenylmethoxy-carbonyl (Fmoc) group can drive self-assembly of short peptide sequences into nanofibrils through π–π stacking of the Fmoc aromatic moieties [38–41]. Previous research has shown that diphenylalanine (FF) can form well-ordered, hollow tubular, and elongated nanofibers in organic solvents [42,43], but cannot form supramolecular structures under biological conditions, without the use of an inorganic solvent. Through the introduction of aromatic capping residues such as the Fmoc-group, this class of SAPs can spontaneously form supramolecular structures using a pH switch method [44]. Ulijn and coworkers identified the self-assembly mechanism whereby these Fmoc-peptide derivatives form antiparallel β-sheets between the pendant peptide resulting in expression of the amino acid side chains on the surface through what is known as π-β interactions (Figure 3.1.D) [45]. The aromatic groups in the Fmoc moiety share electrons to form π–π interactions, interlocking the peptide derivatives. The peptide chains are then brought into close proximity allowing them to interact with each other through the secondary protein structure, antiparallel β-sheets. The combination of several of these assemblies results in the formation of individual nanofibers [40]. Bundles of these nanofibers interact through supramolecular ordering to induce the formation of a highly branched nanofibrous network that presents as a macroscale hydrogel. Importantly, the amino acid side chains of the peptide sequences located on the surface of the nanofibers can provide surface biofunctionality through interactions with cell surface receptors [45,46].

3.5 FABRICATION AND CONTROL OF MECHANICAL PROPERTIES OF PEPTIDE SCAFFOLDS

When designing a new SAP for biological applications, factors that may influence cell behavior must be considered. It has been shown that the matrix stiffness has significant effects on cell fates and behaviors including cell migration, proliferation, and survival (Figure 3.3.C) [47]. For example, mesenchymal stem cells can be directed to differentiate into various cell types (neurogenic, myogenic, or osteogenic) based on the mechanical properties of the cell culture scaffold [48]. This demonstrates the importance of biomechanical cues and how they can affect cell behavior, with matrix elasticity directing stem cell lineage specification [48]. As such, the engineering and design of hydrogel scaffolds with tissue-specific mechanical properties is important both for the stability of the cellular microenvironment and controlling biofunctionality.

The self-assembly of individual peptide building blocks into a nanofibrous network, as seen in various SAP systems, occurs via an external stimulus such as heat, light, pH, temperature, or enzyme. The stimulus modifies the intermolecular interactions, allowing spontaneous self-assembly to take place. Such approaches that fabricate these peptide hydrogels are also related to their mechanical properties and have been extensively investigated. The important factors that determine the mechanical properties of a hydrogel are the average thickness and single fiber mechanical properties, the degree of branching, supramolecular interactions, and the microstructure [49]. Unfortunately, it is often unclear how to control the microstructure of the gels, as well as the degree of cross-linking, although it has been known that gel properties are largely affected by microstructure, and fiber branching can be controlled in related hydrogels by additives.

The properties of hydrogels are also managed by the molecular structure due to the primary assembly being regulated via the intermolecular interactions programmed into the molecule. Clearly the properties of the gels are necessary for specific applications. For example, Stupp and coworkers showed that in a PA, valines increase the stiffness of the gel, while additional alanines decrease the mechanical properties through manipulating the number and position of the two peptides in the amino acid sequence [50]. Molecular control of mechanical properties in

3D artificial peptide scaffolds provides a chemical strategy to control biological phenomena such as stem cell differentiation and cell morphology. For the measurement of mechanical properties, rheology is an effective means of characterizing gels. However, this method is difficult to associate mechanical properties of hydrogels with the molecular structure [49].

3.5.1 pH

Due to the zwitterionic nature of peptides, pH change has been used to induce the protonation/deprotonation of the amino- and carboxyl-groups, with this variation of static electric force causing a related change in the molecular arrangement [51]. For Fmoc-SAPs, the pH of the peptide/water mixture is increased in order to dissolve the Fmoc-peptide by a drop wise addition of concentrated sodium hydroxide, followed by mixing using a vortex until a clear solution is obtained. The pH is then gradually lowered using hydrochloric acid (HCl) so that the individual peptides become optimally charged and interact to spontaneously self-assemble. This usually occurs at a pH several units away from the calculated pK_a value for the molecule. For example, Tang et al. reported that the hydrogels formed by Fmoc-FF at pH 9.0 had higher elastic modulus (G') and viscous modulus (G'') compared to that formed at pH 7.6, due to a more favorable ordering [37]. This method, although a facile route for hydrogel formation, is difficult to control, as the inhomogenous change in pH can mean gelation occurs too quickly due to localized pH changes mostly due to insufficient mixing [52]. This is a relatively uncontrolled method that may lead to the formation of inhomogeneous, turbid hydrogel. In order to surmount this, Adams et al. have developed a system using glucono-δ-lactone (GdL) as a controlled vector of pH change to reproducibly form homogeneous and transparent hydrogels compared with using HCl. This mechanism relies on GdL slowly hydrolyzing in water to produce gluconic acid, homogeneously lowering the pH of the solution [53,54]. Adams et al. also compared the mechanical properties of Fmoc-Y hydrogel formed by using HCl and GdL. Results showed that the gels were much stiffer for the same concentration of peptide using GdL compared to that formed by HCl due to a much more stable and consistent assembly process.

3.5.2 Ionic Strength

Due to the zwitterionic properties of peptides, control of ionic strength has proven to be an effective way to cause hydrogelation and can regulate the final SAP mechanical properties. Yu and coworkers showed that by changing ionic strength, oligopeptide modules such as KVW 10 (acetyl-WKVKVKVKVK-CONH$_2$) can be self-assembled into a hydrogel network [55]. Huang et al. prepared a hydrogel by dissolving the SAP in dimethylsulfoxide (DMSO)/H$_2$O solutions and observed that the addition of Ca^{2+} increased the gel formation speed and also the final gel strength [56]. Ozbas et al. found that with the presence of salt, intramolecular folding could induce the self-assembly of β-hairpin peptides into hydrogels at pH 7.4. They also found that the G' of a peptide hydrogel increased from 100 to 3000 Pa when the ionic strength rose from 20 to 40 mM. A D-form self-assembly peptide system (D-EAK 16) was designed by Luo et al., and results showed that the mechanical properties of D-EAK 16 were improved by the addition of PBS solution with Na$^+$/K$^+$, especially at higher peptide concentration. This may be due to Na$^+$/K$^+$ causing the peptides to self-assemble into short nanofibers at fast speed. However, the stiffness was quite weak ($G' < 20$ Pa at all tested conditions).

3.5.3 Temperature

Hydrogel materials that swell or shrink with changes in temperature are called thermally responsive. As the biological medium is typically 37°C, this is an attractive route to achieve gelation under biological conditions [57]. An example of a thermally responsive SAP is MAX 3 (VKVKVKTKVDPPTKVKTKVKV-NH$_2$), composed of a central tetrapeptide with β-turn propensity flanked by two extended strands. Results showed that the G' of the aqueous solutions of MAX3 was a function of temperature during several cycles of heating and cooling. When temperature was increased, the peptide experienced a unimolecular folding event, providing an amphiphilic β-hairpin and then self-assembled into a hydrogel network. At low concentration (150 μM) and under 80°C, transformation from β-hairpin to β-sheet structure was observed. The folding and unfolding were reversible and dependent on temperature. At room temperature and pH 9, the SAP was unfolded and presented as a low viscosity aqueous solution. When the temperature was increased to 75°C and high concentration (2 wt%), it showed a rigid hydrogel ($G' = 1100$ Pa). Decreasing the temperature to 5°C resulted in a low viscosity (freely flowing) solution as shown by G' values [58]. As the folding and subsequent self-assembly of the β-hairpin are partially controlled by hydrophobic interactions, the particular temperature at which the transition takes place can be tailored by changing the hydrophobic character of the peptide. More hydrophobic peptides should fold and assemble at lower temperatures. The thermal behavior of this class of β-hairpins proves that *de novo* design can be used to predictably construct responsive materials, and that these modifications show the potential for this to occur when a solution of peptides are delivered to the pH range and temperature conditions found in the cellular niche.

3.5.4 Biocatalytic Reactions

The application of enzymes is a promising method to control and direct molecular self-assembly through redefining molecular interactions. As an example, enzymes have been used to induce peptide hydrogelation [59]. Enzyme-catalyzed reactions can form covalent bonds between substrates [60], and have a variety of advantages such as high-specificity and selectivity, mild reaction condition requirements (in aqueous solution, pH 5–8 and biological temperature). The mechanisms for self-assembly are tailored for individual enzymes by tuning the peptide to act as the specific substrate. For example, phosphatase can catalyze bond-cleavage reactions for the production of supramolecular hydrogels through dephosphorylation [61,62] (Figure 3.3.B). TGase can catalyze the formation of isopeptide bonds such as between glutamine and lysine residues. Proteases such as thermolysin have good preference for hydrophobic/aromatic residues on the amine side of the peptide bond and are nonspecific for carboxylic acid residues. They can also catalyze reverse hydrolysis reactions, which may have implications for *in situ* formation of nanofibrous hydrogel scaffolds for cell culture [40]. A reversible system using kinase/phosphatase has been demonstrated. Kinase is used to disrupt the nanofibers formed by the self-assembly of a pentapeptide Nap-FFGEY, and induces a gel–sol transition by then employing phosphatase to restore the self-assembly and form supramolecular hydrogels *in vivo* [63].

Controlling the enzyme concentration can also be an effective way to adjust the mechanical properties of hydrogels by modifying the magnitude of the stimulus. Guilbaud et al. showed that the enzyme concentration does neither affect the self-assembly of FEFK tetra-peptides at a molecular level nor the structure of the fibrillar network formed at the nanoscale using a protease, thermolysin [64]. However, the elastic modulus of the hydrogel increased by almost an order of magnitude when the enzyme concentration was increased from 0.1 to 0.5 mg/mL. This phenomenon is discovered to be due to denser regions forming around the enzymes as they act as a nucleation point. The interactions of these dense regions cause the network to be stronger due to an increased number of entanglements. The utility of this approach has been shown by Yang et al. who used naturally occurring phosphatase to catalyze a β-peptide hydrogels *in vivo* showing excellent biostability (Figure 3.3). It was also found that with the increase of enzyme amount (1.47, 2.94, and 5.88 units/mL), the elastic modulus (300, 900, and 4000 Pa, respectively) of the hydrogels were increased, too, while the gelation times (30, 10, and 2 min) decreased; this effectively shows that the gel properties are related to the number of fibrils and the rate at which they are formed, and this process can be determined by the magnitude of the stimulus [61].

3.5.5 Photopolymerization

The use of light-sensitive compounds, usually called photoinitiators, can be employed as a stimuli responsive trigger to modify a photosensitive intermediate that directly interacts or is attached to the peptides, allowing them to move from solution to an assembled hydrogel [65]. Typically, visible or ultraviolet light is used for photopolymerization. The major advantage of photopolymerization is that hydrogels can be produced *in situ* from aqueous precursors, fast rates (within a second to a few minutes), and functionality at physiological temperature [65,66]. However, some photoinitiators are cytotoxic and large macromolecular precursors are required to drive free radical-mediated cross-linking reactions without disrupting the underlying assembly process. Haines et al. has developed a simple light-activated hydrogelation system using a designed β-turn peptide, MAX7CNB (VKVKVKVKVDPPTKVKXKVKV-NH$_2$) [66]. Under ambient light, a 2 wt% solution was shown to be stable with the peptides in an unfolded state. Irradiation of this solution with light wavelengths between 260 and 360 nm, a photocage was released and peptide folding was triggered to produce amphiphilic β-hairpins and self-assemble into viscoelastic hydrogel material ($G' = 1000$ Pa).

3.6 THE *IN VITRO* AND *IN VIVO* APPLICATIONS OF SELF-ASSEMBLY PEPTIDE SCAFFOLDS

To be effective as a scaffold for regenerative medicine, the supramolecular matrices formed by SAPs must effectively emulate the features of the ECM both morphologically, chemically, and spatially to provide a bioactive and stable 3D microenvironment.

3.7 BIOFUNCTIONALIZATION OF PEPTIDE HYDROGELS

The major challenge in the designing of bioactive biomaterials through molecular self-assembly is to create structurally complex structures on a range of length scales incorporating multiple components that are hierarchically organized, as shown in biological systems [67]. A limitation that must be overcome with SAP scaffolds is the richness of chemistry on the nanoscale must be built upon to replicate this utility to mimic macromolecular higher ordering, either through covalent modification and/or the use of multicomponent systems [68].

3.7.1 Peptide Epitopes Presented by the SAPs

A variety of tissue-specific proteins found in the ECM provide not only physical but also chemical support to cells through initiation of cell pathways via integrin activation [69]. Significant efforts therefore have been made to include functional peptide sequences from these proteins associated with specific functions associated with connective (collagen), nervous (laminin), or mesenchymal (fibronectin) tissues [70,71]. Isolated peptide sequences found in these proteins are known to retain the ability to influence intracellular processes such as adhesion, proliferation, differentiation, and migration [72]. Therefore, significant efforts have been made to include these amino acid sequences in SAPs without disrupting the mechanism of assembly [73]. Unfortunately, taken out of context of the whole protein, the peptide sequences display little or no propensity to assemble and be correctly displayed under physiological conditions. Rodriguez et al. found that through the rational modification of a bioactive peptide sequence with acidic residues, the gel formation pK_a

and consequently, the pH at which gel formed were controllable [69]. Importantly, this modification has no impact on the noncovalent interactions driving the formation of the fibrillar structures. Short peptide epitopes have been incorporated onto PA molecules [34]. RGD, RGDS, IKVAV, and other binding domains have been incorporated into PAs as an approach to improve cellular adhesion. In an effort to create a multicomponent system of mineralized fibers, Stupp et al. incorporated phosphoserine residues into PA scaffolds to promote hydroxyapatite formation *in vitro*. The PA contained the cell adhesion epitope RGDS, and self-assembled into nanofibers and created self-supporting gels under cell culture conditions. These were then shown to nucleate spheroidal nanoparticles of crystalline carbonated hydroxyapatite ~100 nm in diameter. Importantly, the mineralization occurs in the presence of serine or phosphoserine residues in the peptide sequence, not epitaxial relative to the long axis of the nanofibers.

3.7.2 Incorporating Macromolecules to the Scaffold

Instead of taking small sequences and incorporating them to the SAP, research has shown that the fibrillar SAP structures can bind ECM proteins within the scaffold [74,75]. For example, Fmoc-trileucine self assembles to yield fibrils with a high density of leucine that present a highly hydrophobic environment. This was used to bind the ECM protein laminin in a manner analogous to a macromolecule–macromolecule interaction that is found in the ECM [38]. This approach therefore presents as a versatile methodology for the presentation of structural proteins in a potentially biologically relevant manner. More recently, type I collagen was incorporated into PA nanofibers with the IKVAV and YIGSR epitope present on the surface. Homogeneous fibers were formed with 20–30 nm in diameter that underpins the scaffold [76].

PGs are a family of molecules consisting of a core protein covalently linking to one or more glycosaminoglycan sugar chains [77]. Several studies have proven that cell attachment and spreading were improved by modifying biomaterials with PG-binding domains [78–82]. However, Sawyer et al. suggested that when engineering biomimetic strategies for tissue engineering, it is very important to consider the unique surface properties of specific biomaterial integrates with endogenous processes such as protein adsorption allows functional presentation [83]. Recently, a bioactive hierarchically structured membrane formed with PAs (LRK-KLGKA) and hyaluronic acid [67].

3.8 SAP SCAFFOLDS AS A SUPPORT FOR 3D CELL CULTURE

3.8.1 Alpha Helical

A 28-residue α-helical peptide was developed by Woolfson and coworkers with tryptophan used to stabilize the gel in cell culture media at 37°C. The resultant hydrogel can support bone growth and differentiation of rat adrenal pheochromocytoma cells in a continuous culture [23]. An alanine-rich amphiphilic peptide-containing RGD cell adhesion motif A_6RGD was used to research on fibroblast attachment. Results showed that its self-assembly was dependent on the peptide concentration in water as a nanofibril or vesicle structure was available. It showed that human cornea stromal fibroblasts (hCSFs) proliferation was enhanced significantly at 0.1–1.0 wt% A_6RGD solutions compared with the control [84].

3.8.2 β-Sheets

One of the predominant systems to utilize β sheets are the RAD16 based peptide scaffolds. These have been used to support a variety of applications, including osteoblast proliferation and differentiation [85], axon differentiation [86], keratinocyte differentiation [87], positive effects on neural regeneration following injury to the central nervous system [88] and, in an important demonstration of the long-term effectiveness of these materials, have encouraged the deposition of new ECM matrix proteins [89]. Recently, it was used as a 3D model for ovarian cancer cells, and results showed higher resistance to anticancer drugs compared to cells grown on a 2D culture plate, it is possible to use as an anticancer drug screening and *in vitro* investigation of tumor biology [90]. Both of RAD16-II and EAK16-II were able to support the attachment and growth of a number of mammalian cells such as fibroblasts and keratinocytes [91]. While another β-sheet structured octapeptide FEFEFKFK scaffold was used for culturing bovine chondrocytes and its biocompatibility was assessed. 2D cell culture demonstrated collagen type I deposition, the support of cell viability, and the retention of cell morphology in 3D cell culture. This showed that the scaffold may serve as a template for cartilage tissue engineering [92].

3.8.3 β-Turns

The β-hairpin formed peptide MAX1 has shown to be cytocompatible, and capable of supporting the cellular attachment of NIH3T3 cells, expanding red blood cells, and fibroblast proliferation [26]. Progenitor osteoblast cells MG63 cultured in MAX1 remained evenly distributed and viable within the hydrogel during delivery by syringe injection, indicating the potential of using this type of hydrogel as an injectable cell scaffold [28]. While MAX1 is not ideal for cell encapsulation due to slow gelation speed, MAX8 can encapsulate cells with a homogeneous distribution (Figure 3.1.C) [27]. Importantly, however, both MAX1 and MAX8 have not demonstrated a significant macrophage response *in vivo* [93].

3.8.4 Unnatural Amino Acids

Unnatural amino acids are of interest for cell culture as they, if proven to be noncytotoxic, produce stable materials that are not broken down by enzymes produced in the body. SMMC7721 cells were cultured on the protease-resistant D-form peptide sequence D-EAK16. The results showed high cell viability and low-level cell apoptosis for several weeks with no significant difference of viability as compared with the L-form peptide [94]. Tekinay and coworkers incorporated charged groups such as sulfonate, carboxylate, and hydroxyl on SAP amphiphile nanofibers in order to mimic ECM chemically and structurally [95]. The differentiation of ATDC5 chondrogenic cells was evaluated and results showed that although peptide nanofiber systems were different to each other due to different chemical composition, ATDC5 cells aggregated quickly and formed cartilage-like nodules in insulin-free medium and deposited sulfated glycosaminoglycans when cultured on these functional peptide scaffolds.

3.8.5 Short Peptide Derivatives

Short peptide sequences (1–3 residues) have been shown to be highly effective at forming stable nanostructures. However, due to their limited size, it is a challenge to introduce peptide sequences that are bioactive. Chemical functionality arising from the limited amino acid side chains available to this approach (such as $-NH_2$, $-COOH$, and $-OH$) has been used to study if they can improve the scaffold compatibility with different cell types. To do this, Fmoc-amino acids such as Fmoc-G (R = $(CH_2)_4NH_2$), Fmoc-D (R= CH_2COOH), and Fmoc-S (R= CH_2OH) were mixed with the structurally effective, but biologically inert Fmoc-FF to form hydrogels (Figure 3.2.A) [52]. These hydrogels formed antiparallel β-sheet structure and were all softer than Fmoc-FF with a G' from 502 Pa (Fmoc-FF/D) to 21.2 kPa (Fmoc-FF), suggesting a disruption in structure. Cell culture experiments showed that bovine chondrocytes survived in all the four hydrogels and in which Fmoc-FF/S and Fmoc-FF/D could support human dermal fibroblasts (HDF), and Fmoc-FF/S was the only gel type that supported viability for all three cell types tested (bovine chondrocytes, mouse 3T3 fibroblasts, and HDF). In an extension of this work, Ulijn and coworkers also made a mixture of two aromatic peptides, Fmoc-RGD and Fmoc-FF and found that with 10–30% combining, Fmoc-FF/RGD gels ($G' = 4$–10 kPa) are stiffer than

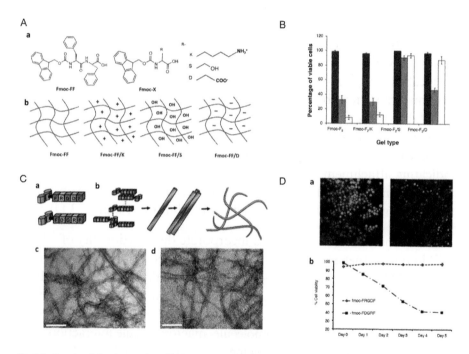

Fig. 3.2. Functionalizing the peptide scaffold. (A) Functionalizing minimalist amino acid sequences by mixing in charged residues. Chemical structures (a) and a representation (b) of the four hydrogels. Adapted with permission from Ref. [52]. Copyright (2009) Elsevier. (B) Comparison of Fmoc-FF, Fmoc-FF/K (1:1), Fmoc-FF/S (1:1), Fmoc-FF/D (1:1) in 2D culture of chondrocytes (black), 3T3 fibroblasts (gray), and HDFs (white). The introduction of Fmoc-serine and Fmoc aspartic acid were far more effective in cell culture than the aromatic Fmoc-F and the basic lysine. Adapted with permission from Ref. [52]. Copyright (2009) Elsevier. (C) (a) Schematic representation of the SAPs Fmoc-FRGDF and the control sequence Fmoc-FDGRF. (b) π-Stacking (dark gray [red in the web version) arrow] and β-sheet interactions [light gray (orange in the web version) arrow] account for the formation of nanofibrils, which self-assemble to form supramolecular matrix. (c–d) Transmission electronic microscope (TEM) images of the two peptide hydrogels (C, Fmoc-FRGDF; D, Fmoc-FDGRF) underpinned by nanofibers (scale bar 50 nm). Adapted with permission from Ref. [97]. Copyright (2014) Wiley Periodicals, Inc. (D) Despite being mechanically and morphologically similar, the DGR hydrogel is far less suitable than the fibronectin-derived RGD peptide. (a) Live dead stain showing high cell viability in the RGD gel [light gray (green in the web version) cells] and fewer live cells and many dead cells in the DGR hydrogel. (b) Shows how this varies over a period of days in culture. Adapted with permission from Ref. [97]. Copyright (2014) Wiley Periodicals, Inc.

100% Fmoc-FF ($G'\approx1.9$ kPa) [96]. The gel with 20% Fmoc-RGD incorporated had the maximum G' of 10 kPa, while when Fmoc-RGD content was over 30%, the gel was softer than 100% Fmoc-FF. These results demonstrate that there are complicated relationships between molecular composition, supramolecular arrangement, and hydrogel properties. 3D cell culture experiments on Fmoc-FF/RGD scaffold (30% Fmoc-RGD) showed that presentation of the RGD sequence within the Fmoc-FF system, promoted adhesion and subsequently the spreading and proliferation of cells. Such mixed-peptide systems may offer an alternative, economical approach, particularly when one peptide epitope within the system (in this case, Fmoc-RGD) cannot form supramolecular structures at physiological pH [59]. More recently, Williams and coworkers inserted RGD sequence into Fmoc-FRGDF, which formed an ~10 nm nanotube with the high density of RGD present on the surface of the nanotube through π–β assembly (Figure 3.2.C) [97]. The 3D cell of human mammary fibroblast cells show high levels of viability in the Fmoc-FRGDF system, but when cultured in a structurally and chemically similar system, Fmoc-FDGRF, reduced viability is observed. Furthermore, in the RGD system, actin staining showed the formation of filopodia indicating attachment, whereas in the DGR system, only those cells directly in contact with the tissue culture plastic (TCP) base were observed to spread and display flattened, satellite morphology (Figure 3.3.D) [97].

Efforts have also been made to incorporate the functionality of whole ECM proteins that are biologically active through supramolecular interactions [74,75]. A tripeptide, Fmoc-L$_3$, was enzymatically catalyzed to produce fibers with a high density of leucine ensuring a hydrophobic surface was presented by the fibrils. This was used to bind the ECM protein laminin in a manner analogous to a macromolecule–macromolecule interaction [38]. More recently, type I collagen was incorporated into a PA nanofiber with the IKVAV and YIGSR epitope present on the surface. Homogeneous fibers were formed with 20–30 nm in diameter that underpin the scaffold. 3D cell culture with two major neuronal subtypes of cerebellar cortex, granule cells, and Purkinje cells showed distinct response to the change of epitope concentration. Due to the ability to modulate neuron survival and maturation by easy manipulation of epitope density, this work offers a versatile method to research the contribution of ECM to neuron development and the design of optimal neuronal scaffold biomaterials [76].

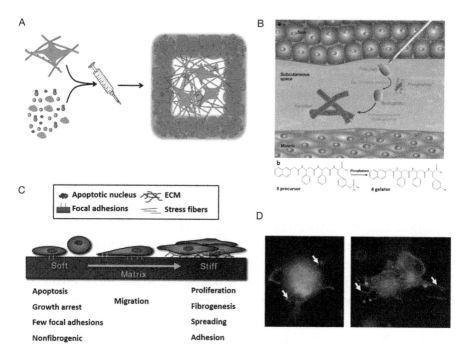

Fig. 3.3. Triggering SAP scaffolds for 3D cell culture. (A) Schematic illustrating the general methodology – either cells are cultured on a preformed SAP hydrogel, or the building blocks and cells are coassembled. Once in vivo, *the material supports the growth of the transplanted cells, fills a void, and interacts with the surrounding tissue.* Adapted with permission from Ref. [11]. Copyright (2012). *(B) Using the host enzymes to trigger assembly of β-amino acids.* Adapted with permission from Ref. [61]. Copyright (2007) WILEY-VCH Verlag GmbH & Co. KGaA, Weinheim. *(C) Effects of ECM mechanical properties on cell behavior. Illustration of the general changes in cell behavior observed with the increase of matrix stiffness. Generally cells are rounded and minimally attached to soft matrices as well as growth arrested (or minimally proliferative) and liable to apoptosis. While cells cultured on stiff matrices are prone to proliferate and be fibrogenic with a spread phenotype, increased numbers of integrin/ ECM bonds, and in the case of fibroblasts, stress fibers. Cells migrate from soft to stiff area of the matrix and may be most motile at intermediate stiffness.* Adapted with permission from Ref. [47]. Copy (2008) American Association for the Study of Liver Diseases. *(D) Fibroblast cells cultured on a soft 3D scaffold and a 2D hard surface. The arrows indicate stress fibers.* Adapted with permission from Ref. [97]. Copyright (2014) Wiley Periodicals, Inc.

3.9 UTILIZING SELF-ASSEMBLY PEPTIDE SCAFFOLD AS CELL THERAPY *IN VIVO*

The ultimate goal for cell therapy is to transfer cells to the human body, so the *in vivo* study of SAP scaffold is essential for this purpose. One of the major reasons is *in vitro* cell culture has little relationship to the dynamic natural cellular environment For *in vivo* applications, the self-assembly process occurs either *in vitro* or the peptide in solution mixed with cells and signals assembling *in vivo* (Figure 3.3). They must be transferred by a minimally invasive delivery vector, such as microinjection. Then the peptide scaffolds communicate with the implanted cells, mechanically supporting the host tissue, attenuating the immune response, and/or allowing the migration into the scaffold of native cells (Figure 3.3) [11]. A range of peptide scaffolds have been used *in vivo* and shown significant potential as adjuvant scaffolds for tissue engineering.

3.9.1 Myocardium

Heart failure is one of the major causes of death in Western countries and is expected to become a global epidemic within the twenty-first century. SAP hydrogel scaffolds have been investigated to improve the survival of implanted cells for myocardial restoration. Hydrogels made of RAD16 peptides are stable, illicit no inflammation after implantation and support the migration of cells into the myocardium [98]. These SAP nanofibers were then used to deliver binding to platelet-derived growth factor-BB (PDGF-BB). Results show that the nanofibers maintained delivery of PDGF-BB for 14 days with enhanced cell viability and improved systolic function after myocardial infarction [99]. Again, using SAP scaffolds to deliver vascular endothelial growth factor (VEGF) has been used for the treatment of postinfarction neovascularization in rats. Results showed that SAP/VEGF injection dramatically improved arteriogenesis and cardiac performance in the 28 days after myocardial infarction (Figure 3.4.A) [100].

3.9.2 Bone Regeneration

The use of SAP scaffolds to deliver growth factors has a significant utility in bone repair. Bone defects are common problems and they are difficult to cure with current therapies. Therapeutic proteins such as morphogenetic proteins, transforming growth factor-β, and basic fibroblast growth factor need to be delivered *in vivo* so that they can induce bone formation. However, the use of these proteins alone requires large amounts of protein due to their short half-lives *in vivo*. A 3D hybrid SAP/collagen scaffold was engineered to release growth factor sustainably. After implanting the PA hybrid peptide scaffold into the backs of rats, homogeneous bone formation was morphologically observed throughout the hybrid scaffolds. The results showed that the incorporation of basic fibroblast growth factor combined in a self-assembled PA nanofiber hybrid scaffold is a promising method to promote bone formation without requiring autologous bone grafts (Figure 3.4.B) [101].

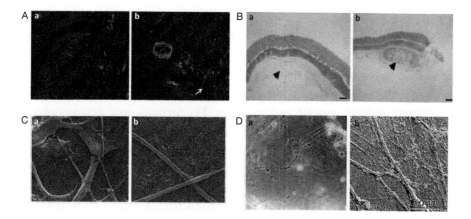

Fig. 3.4. *SAP scaffolds for regenerative medicine.* *(A) Nanofibrous SAP matrices encourage the formation of new arterioles in the myocardium that has suffered an infarction (a) shows the control, whereas in the presence of the SAP, smooth muscle cells can be clearly seen [SM22a; white (green in the web version)], cardiomyocytes [light gray (red in the web version)], and nuclei [dark gray (blue in the web version)] at the infarct border zone from each group.* Adapted with permission from Ref. [100]. Copyright (2012) American Association for the Advancement of Science. *(B) Histologic cross-sections of ectopically formed bone 4 weeks after implantation of collagen sponges reinforced with incorporation of PGA fiber, incorporated with PA (a), basic fibroblast growth factor (bFGF) (b), and self-assembled PA nanofibers (hybrid scaffold).* Adapted with permission from Ref. [101]. Copyright (2007) Mary Ann Liebert, Inc. *(C) Heparin-like signals encourages angiogenesis – electron micrographs of endothelial sprouting and tube formation on heparin-mimetic PA matrix. (b) Endothelial cells interact with the fibrous network and form polygonal structures. (a) Endothelial sproutings are clearly observed interacting with the PA matrix.* Adapted with permission from Ref. [104]. Copyright (2011) American Chemical Society. *(D) Primary mouse cerebellar granule neurons project extensive neurites on scaffolds. The confocal images show bright discrete light gray (green in the web version) labeling indicative of synaptically active membranes.* Adapted with permission from Ref. [105].

3.9.3 Cartilage Repair

SAP scaffolds have also been applied in the cartilage regeneration. A tripeptide KLD solution was mixed with bone marrow stem cells (BMSC), growth factors, and an anti-inflammatory agent. This solution was injected into skeletally mature rabbit model with a full-sized cartilage defect, whereupon the SAP assembled in response to the physiological environment, filling the defect and encapsulating the cells [102]. The results show that collagen II and Safranin-O production were improved, and the defect repaired significantly, allowing functional recovery. Additionally, the mechanical properties of peptide gels can have a significant therapeutic benefit. An 11-mer peptide, which mimics hyaluronic acid, has been found to be an effective lubricant [103]. It can reduce friction and induce lubrication effectively, which shows that SAPs may be used as an alternate therapeutic lubricant for osteoarthritis.

3.9.4 Angiogenesis

Angiogenesis, or revascularization, is one of the most important processes required for functional tissue formation in order to ensure a good supply of nutrients, oxygen, and the removal of toxic metabolites. Guler and co-workers synthesized several heparin-mimetic and binding PA molecules to give functionalized nanofibrous scaffolds. The formation of new vessel was induced *in vitro* and also in Sprague-Dawley rats; robust vascularization was observed in rats due to the bioactive interactions between the nanofibers and the bound growth factors. These heparin-mimetic peptide nanofibers provide new views for angiogenesis and tissue regeneration without resort to synthetic heparin and exogenous growth factors (Figure 3.4.C) [104].

3.9.5 Neural Regeneration

SAP scaffolds show potential for the regeneration of neuronal cells, as the soft hydrogel materials are similar in morphology and mechanical properties to the brain. Zhang and coworkers developed RAD16-I and RAD16-II. Results showed that the scaffolds were able to support neuronal cell attachment and differentiation. Furthermore, extensive neurite outgrowths followed the profiles of the scaffold and were able to form active synaptic connections in rats, showing the potential for guided regeneration. Importantly, these peptides did not induce a noticeable immune response *in vivo* (Figure 3.4.D) [105]. In an extension of this work, Cheng et al. attached laminin-derived IKVAV motif on the C-terminal of the assembling fragment $RADA_{16}$ hydrogel in order to provide a functionalized scaffold to deliver encapsulated neural stem cells (NSCs) to defects within the injured brain [68]. The hydrogel formed bilayer β-sheet structure with mechanical properties similar to brain tissue. The results (*in vitro*) have shown that the extended peptide sequence can be used to guide NSCs adhesion and differentiation. When trials (*in vivo*) were conducted on a rat brain surgery model, results showed that the injected peptide solution rapidly formed a 3D hydrogel *in situ*, filling the injury cavity. As such, void formation was reduced and enhanced cell viability as well as a decrease in recruitment of glial astrocytes to the site of injury was observed. Although both $RADA_{16}$ and $RADA_{16}$-IKVAV hydrogels supported transplanted NSCs survival, growth, and distribution, NSCs tended to differentiate into mature neuronal cells in the $RADA_{16}$-IKVAV scaffold, while they were more likely to become astroglia cells in the nonfunctionalized $RADA_{16}$ matrices, highlighting the effect of the epitope, IKVAV on NSCs performance *in vivo*.

3.10 FUTURE PERSPECTIVES

The chemically and morphologically rich scaffolds formed from the assembly of peptides and peptide-containing materials are excellent candidates for biomaterials. This is mostly due to the structural modifications, inherent biocompatibility, and biodegradability of peptide sequences [11]. Native tissues have a varied range of morphologies and physiological properties; these features affect cell survival, migration, and proliferation [47]. These structures form via supramolecular assembly. Therefore, by taking inspiration from these processes, researchers can hope to achieve application-specific control over the endogenous cell behaviors through effective biomimicry of the biochemical and mechanical microenvironment. They *in situ* triggered self-assembly that occurs under mild, physiological conditions enables easy incorporation of cells during hydrogel formation and therefore, the potential for use as a minimally invasive injectable therapy [106]. Recently, a greater understanding of the structural design and self-assembly processes has allowed specific modifications to be made to these materials for the specific requirements for applications in regenerative medicine. Additionally, while the use of SAPs have largely been limited to soft tissue engineering, the development of hybrid systems may provide an interesting route in extending their applications. However, it has become clear that the next generation of these materials must more accurately mimic the natural ECM through the introduction of features over several length scales, as biology is ordered hierarchically; they must also become more dynamic, as they need to be able to change over time, and in response to the developmental requirements of the systems. In conclusion, the complete potential of SAP scaffolds for cell therapy and tissue engineering has yet to be fully realized; their multifunctionality in producing tailored biologically active landscape warrants further research.

REFERENCES

[1] Cell therapy, Nature 392 (1998) 18–24.

[2] D.E. Discher, D.J. Mooney, P.W. Zandstra, Growth factors, matrices, and forces combine and control stem cells, Science 324 (2009) 1673–1677.

[3] M.R. Ghadiri, Self-assembled nanoscale tubular ensembles, Adv Mater 7 (1995) 675–677.

[4] M.P. Lutolf, J.A. Hubbell, Synthetic biomaterials as instructive extracellular microenvironments for morphogenesis in tissue engineering, Nat Biotechnol 23 (2005) 47–55.

[5] S.H. Kim, J. Turnbull, S. Guimond, Extracellular matrix and cell signalling: the dynamic cooperation of integrin, proteoglycan and growth factor receptor, J Endocrinol 209 (2011) 139–151.

[6] R. Ayala, C. Zhang, D. Yang, Y. Hwang, A. Aung, S.S. Shroff, et al. Engineering the cell–material interface for controlling stem cell adhesion, migration, and differentiation, Biomaterials 32 (2011) 3700–3711.

[7] M.W. Tibbitt, K.S. Anseth, Hydrogels as extracellular matrix mimics for 3D cell culture, Biotechnol Bioeng 103 (2009) 655–663.

[8] S. Zhang, X. Zhao, Design of molecular biological materials using peptide motifs, J Mater Chem 14 (2004) 2082–2086.

[9] S. Zhang, Building from the bottom up, Mater Today 6 (2003) 20–27.

[10] B. Grzybowski, Self-assembly at all scales, Science 295 (2002) 2418–2421.

[11] R.J. Williams, Self-assembled peptides: characterisation and *in vivo* response, Biointerphases 7 (2012) 1–14.

[12] P. Bures, W. Leobandung, H. Ichikawa, Hydrogels in pharmaceutical formulations, Eur J Pharmaceut Biopharmaceut 50 (2000) 27–46.

[13] C. Frantz, K.M. Stewart, V.M. Weaver, The extracellular matrix at a glance, J Cell Sci 123 (2010) 4195–4200.

[14] R.V. Ulijn, A.M. Smith, Designing peptide based nanomaterials, Chem Soc Rev 37 (2008) 664–675.

[15] K. Rajagopal, J.P. Schneider, Self-assembling peptides and proteins for nanotechnological applications, Curr Opin Struc Biol 14 (2004) 480–486.

[16] E.C. Wu, S. Zhang, C.A.E. Hauser, Self-assembling peptides as cell-interactive scaffolds, Adv Funct Mater 22 (2012) 456–468.

[17] R.J. Mart, R.D. Osborne, M.M. Stevens, R.V. Ulijn, Peptide-based stimuli-responsive biomaterials, Soft Matter 2 (2006) 822–835.

[18] P. Burkhard, J. Stetefeld, S.V. Strelkov, Coiled coils: a highly versatile protein folding motif, Trends Cell Biol 11 (2001) 82–88.

[19] A.M. Smith, E.F. Banwell, W.R. Edwards, M.J. Pandya, D.N. Woolfson, Engineering increased stability into self-assembled protein fibers, Adv Funct Mater 16 (2006) 1022–1030.

[20] M.G. Ryadnov, D.N. Woolfson, Engineering the morphology of a self-assembling protein fibre, Nat Mater 2 (2003) 329–332.

[21] D. Papapostolou, A.M. Smith, E.D.T. Atkins, S.J. Oliver, M.G. Ryadnov, L.C. Serpell, et al. Engineering nanoscale order into a designed protein fiber, Proc Natl Acad Sci 104 (2007) 10853–10858.

[22] A.M. Smith, S.F.A. Acquah, N. Bone, H.W. Kroto, M.G. Ryadnov, M.S.P. Stevens, et al. Polar assembly in a designed protein fiber, Angew Chem Int Ed 44 (2005) 325–328.

[23] E.F. Banwell, E.S. Abelardo, D.J. Adams, M.A. Birchall, A. Corrigan, A.M. Donald, et al. Rational design and application of responsive α-helical peptide hydrogels, Nat Mater 8 (2009) 596–600.

[24] A. Aggeli, I.A. Nyrkova, M. Bell, R. Harding, L. Carrick, T.C.B. McLeish, et al. Hierarchical self-assembly of chiral rod-like molecules as a model for peptide β-sheet tapes, ribbons, fibrils, and fibers, Proc Natl Acad Sci 98 (2001) 11857–11862.

[25] S. Zhang, T. Holmes, C. Lockshin, A. Rich, Spontaneous assembly of a self-complementary oligopeptide to form a stable macroscopic membrane, Proc Natl Acad Sci USA 90 (1993) 3334–3338.

[26] J.K. Kretsinger, L.A. Haines, B. Ozbas, D.J. Pochan, J.P. Schneider, Cytocompatibility of self-assembled β-hairpin peptide hydrogel surfaces, Biomaterials 26 (2005) 5177–5186.

[27] L. Haines-Butterick, K. Rajagopal, M. Branco, D. Salick, R. Rughani, M. Pilarz, et al. Controlling hydrogelation kinetics by peptide design for three-dimensional encapsulation and injectable delivery of cells, Proc Natl Acad Sci 104 (2007) 7791–7796.

[28] C. Yan, A. Altunbas, T. Yucel, R.P. Nagarkar, J.P. Schneider, D.J. Pochan, Injectable solid hydrogel: mechanism of shear-thinning and immediate recovery of injectable β-hairpin peptide hydrogels, Soft Matter 6 (2010) 5143–5156.

[29] S. Matsumura, S. Uemura, H. Mihara, Fabrication of nanofibers with uniform morphology by self-assembly of designed peptides, Chem Eur J 10 (2004) 2789–2794.

[30] T.D. Sargeant, C. Aparicio, J.E. Goldberger, H. Cui, S.I. Stupp, Mineralization of peptide amphiphile nanofibers and its effect on the differentiation of human mesenchymal stem cells, Acta Biomater 8 (2012) 2456–2465.

[31] D.M. Marini, W. Hwang, D.A. Lauffenburger, S. Zhang, R.D. Kamm, Left-handed helical ribbon intermediates in the self-assembly of a β-sheet peptide, Nano Lett 2 (2002) 295–299.

[32] J.D. Hartgerink, E. Beniash, S.I. Stupp, Self-assembly and mineralization of peptide-amphiphile nanofibers, Science 294 (2001) 1684–1688.

[33] J.D. Hartgerink, E. Beniash, S.I. Stupp, Peptide-amphiphile nanofibers: a versatile scaffold for the preparation of self-assembling materials, Proc Natl Acad Sci 99 (2002) 5133–5138.

[34] S.E. Paramonov, H.-W. Jun, J.D. Hartgerink, Self-assembly of peptide – amphiphile nanofibers: the roles of hydrogen bonding and amphiphilic packing, J Am Chem Soc 128 (2006) 7291–7298.

[35] C.A.E. Hauser, R. Deng, A. Mishra, Y. Loo, U. Khoe, F. Zhuang, et al. Natural tri- to hexapeptides self-assemble in water to amyloid β-type fiber aggregates by unexpected α-helical intermediate structures, Proc Natl Acad Sci 108 (2011) 1361–1366.

[36] A. Mishra, Y. Loo, R. Deng, Y.J. Chuah, H.T. Hee, J.Y. Ying, et al. Ultrasmall natural peptides self-assemble to strong temperature-resistant helical fibers in scaffolds suitable for tissue engineering, Nano Today 6 (2011) 232–239.

[37] C. Tang, A.M. Smith, R.F. Collins, R.V. Ulijn, A. Saiani, Fmoc-diphenylalanine self-assembly mechanism induces apparent pK_a shifts, Langmuir 25 (2009) 9447–9453.

[38] R.J. Williams, T.E. Hall, V. Glattauer, J. White, P.J. Pasic, A.B. Sorensen, et al. The *in vivo* performance of an enzyme-assisted self-assembled peptide/protein hydrogel, Biomaterials 32 (2011) 5304–5310.

[39] Z. Yang, B. Xu, A simple visual assay based on small molecule hydrogels for detecting inhibitors of enzymes, Chem Commun 2004 (2004) 2424–2425.

[40] S. Toledano, R.J. Williams, V. Jayawarna, R.V. Ulijn, Enzyme-triggered self-assembly of peptide hydrogels via reversed hydrolysis, J Am Chem Soc 128 (2006) 1070–1071.

[41] A. Mahler, M. Reches, M. Rechter, S. Cohen, E. Gazit, Rigid, self-assembled hydrogel composed of a modified aromatic dipeptide, Adv Mater 18 (2006) 1365–1370.

[42] M. Reches, E. Gazit, Casting metal nanowires within discrete self-assembled peptide nanotubes, Science 300 (2003) 625–627.

[43] M. Reches, E. Gazit, Controlled patterning of aligned self-assembled peptide nanotubes, Nat Nano 1 (2006) 195–200.

[44] Y. Zhang, H. Gu, Z. Yang, B. Xu, Supramolecular hydrogels respond to ligand-receptor interaction, J Am Chem Soc 125 (2003) 13680–13681.

[45] A.M. Smith, R.J. Williams, C. Tang, P. Coppo, R.F. Collins, M.L. Turner, et al. Fmoc-diphenylalanine self assembles to a hydrogel via a novel architecture based on π–π interlocked β-sheets, Adv Mater 20 (2008) 37–41.

[46] H. Xu, A.K. Das, M. Horie, M.S. Shaik, A.M. Smith, Y. Luo, et al. An investigation of the conductivity of peptide nanotube networks prepared by enzyme-triggered self-assembly, Nanoscale 2 (2010) 960–966.

[47] R.G. Wells, The role of matrix stiffness in regulating cell behavior, Hepatology 47 (2008) 1394–1400.

[48] A.J. Engler, S. Sen, H.L. Sweeney, D.E. Discher, Matrix elasticity directs stem cell lineage specification, Cell 126 (2006) 677–689.

[49] J. Raeburn, D.J. Adams, The importance of the self-assembly process to control mechanical properties of low molecular weight hydrogels, Chem Soc Rev 42 (2013).

[50] E.T. Pashuck, H. Cui, S.I. Stupp, Tuning supramolecular rigidity of peptide fibers through molecular structure, J Am Chem Soc 132 (2010) 6041–6046.

[51] S.Y. Qin, S.S. Xu, R.X. Zhuo, X.Z. Zhang, Morphology transformation via pH-triggered self-assembly of peptides, Langmuir 28 (2012) 2083–2090.

[52] V. Jayawarna, S.M. Richardson, A.R. Hirst, N.W. Hodson, A. Saiani, J.E. Gough, et al. Introducing chemical functionality in Fmoc-peptide gels for cell culture, Acta Biomater 5 (2009) 934–943.

[53] D.J. Adams, M.F. Butler, W.J. Frith, M. Kirkland, L. Mullen, P. Sanderson, A new method for maintaining homogeneity during liquid–hydrogel transitions using low molecular weight hydrogelators, Soft Matter 5 (2009) 1856–1862.

[54] L. Chen, K. Morris, A. Laybourn, D. Elias, M.R. Hicks, A. Rodger, et al. Self-assembly mechanism for a naphthalene-dipeptide leading to hydrogelation, Langmuir 26 (2010) 5232–5242.

[55] S. Ramachandran, P. Flynn, Y. Tseng, Y.B. Yu, Electrostatically controlled hydrogelation of oligopeptides and protein entrapment, Chem Mater 17 (2005) 6583–6588.

[56] H. Huang, A.I. Herrera, Z. Luo, O. Prakash, X.S. Sun, Structural transformation and physical properties of a hydrogel-forming peptide studied by NMR, transmission electron microscopy, and dynamic rheometer, Biophys J 103 (2012) 979–988.

[57] C. Wang, R.J. Stewart, J. Kopeček, Hybrid hydrogels assembled from synthetic polymers and coiled-coil protein domains, Nature 397 (1999) 417–420.

[58] D.J. Pochan, J.P. Schneider, J. Kretsinger, B. Ozbas, K. Rajagopal, L. Haines, Thermally reversible hydrogels via intramolecular folding and consequent self-assembly of a de novo designed peptide, J Am Chem Soc 125 (2003) 11802–11803.

[59] R.J. Williams, R.J. Mart, R.V. Ulijn, Exploiting biocatalysis in peptide self-assembly, Biopolymers 94 (2010) 107–117.

[60] J.J. Sperinde, L.G. Griffith, Synthesis and characterization of enzymatically-cross-linked poly(ethylene glycol) hydrogels, Macromolecules 30 (1997) 5255–5264.

[61] Z. Yang, G. Liang, M. Ma, Y. Gao, B. Xu, In vitro and in vivo enzymatic formation of supramolecular hydrogels based on self-assembled nanofibers of a β-amino acid derivative, Small 3 (2007) 558–562.

[62] Z. Yang, B. Xu, Using enzymes to control molecular hydrogelation, Adv Mater 18 (2006) 3043–3046.

[63] Z. Yang, G. Liang, L. Wang, B. Xu, Using a kinase/phosphatase switch to regulate a supramolecular hydrogel and forming the supramolecular hydrogel in vivo, J Am Chem Soc 128 (2006) 3038–3043.

[64] J.B. Guilbaud, C. Rochas, A.F. Miller, A. Saiani, Effect of enzyme concentration of the morphology and properties of enzymatically triggered peptide hydrogels, Biomacromolecules 14 (2013) 1403–1411.

[65] K.T. Nguyen, J.L. West, Photopolymerizable hydrogels for tissue engineering applications, Biomaterials 23 (2002) 4307–4314.

[66] L.A. Haines, K. Rajagopal, B. Ozbas, D.A. Salick, D.J. Pochan, J.P. Schneider, Light-activated hydrogel formation via the triggered folding and self-assembly of a designed peptide, J Am Chemi Soc 127 (2005) 17025–17029.

[67] L.W. Chow, R. Bitton, M.J. Webber, D. Carvajal, K.R. Shull, A.K. Sharma, et al. A bioactive self-assembled membrane to promote angiogenesis, Biomaterials 32 (2011) 1574–1582.

[68] T.-Y. Cheng, M.-H. Chen, W.-H. Chang, M.-Y. Huang, T.-W. Wang, Neural stem cells encapsulated in a functionalized self-assembling peptide hydrogel for brain tissue engineering, Biomaterials 34 (2013) 2005–2016.

[69] A.L. Rodriguez, C.L. Parish, D.R. Nisbet, R.J. Williams, Tuning the amino acid sequence of minimalist peptides to present biological signals via charge neutralised self assembly, Soft Matter 9 (2013).

[70] G.A. Silva, C. Czeisler, K.L. Niece, E. Beniash, D.A. Harrington, J.A. Kessler, et al. Selective differentiation of neural progenitor cells by high-epitope density nanofibers, Science 303 (2004) 1352–1355.

[71] H.S. Koh, T. Yong, C.K. Chan, S. Ramakrishna, Enhancement of neurite outgrowth using nano-structured scaffolds coupled with laminin, Biomaterials 29 (2008) 3574–3582.

[72] D.R. Zimmermann, M.T. Dours-Zimmermann, Extracellular matrix of the central nervous system: from neglect to challenge, Histochem Cell Biol 130 (2008) 635–653.

[73] N.R. Lee, C.J. Bowerman, B.L. Nilsson, Effects of varied sequence pattern on the self-assembly of amphipathic peptides, Biomacromolecules (2013).

[74] S. Banta, I.R. Wheeldon, M. Blenner, Protein engineering in the development of functional hydrogels, Ann Rev Biomed Eng 12 (2010) 167–186.

[75] Y.S. Jo, S.C. Rizzi, M. Ehrbar, F.E. Weber, J.A. Hubbell, M.P. Lutolf, Biomimetic PEG hydrogels crosslinked with minimal plasmin-sensitive tri-amino acid peptides, J Biomed Mater Res A 93 (2010) 870–877.

[76] S. Sur, E.T. Pashuck, M.O. Guler, M. Ito, S.I. Stupp, T. Launey, A hybrid nanofiber matrix to control the survival and maturation of brain neurons, Biomaterials 33 (2012) 545–555.

[77] S. Ohshika, Y. Ishibashi, A. Kon, T. Kusumi, H. Kijima, S. Toh, Potential of exogenous cartilage proteoglycan as a new material for cartilage regeneration, Int Orthop 36 (2012) 869–877.

[78] M. Dettin, M.T. Conconi, R. Gambaretto, A. Bagno, C. Di Bello, A.M. Menti, et al. Effect of synthetic peptides on osteoblast adhesion, Biomaterials 26 (2005) 4507–4515.

[79] S. Sagnella, E. Anderson, N. Sanabria, R.E. Marchant, K. Kottke-Marchant, Human endothelial cell interaction with biomimetic surfactant polymers containing Peptide ligands from the heparin binding domain of fibronectin, Tissue Eng 11 (2005) 226–236.

[80] M. Dettin, M.T. Conconi, R. Gambaretto, A. Pasquato, M. Folin, C. Di Bello, et al. Novel osteoblast-adhesive peptides for dental/orthopedic biomaterials, J Biomed Mater Res 60 (2002) 466–471.

[81] A. Rezania, K.E. Healy, Biomimetic peptide surfaces that regulate adhesion, spreading, cytoskeletal organization, and mineralization of the matrix deposited by osteoblast-like cells, Biotechnol Progr 15 (1999) 19–32.

[82] K.C. Dee, T.T. Andersen, R. Bizios, Design and function of novel osteoblast-adhesive peptides for chemical modification of biomaterials, J Biomed Mater Res 40 (1998) 371–377.

[83] A.A. Sawyer, K.M. Hennessy, S.L. Bellis, The effect of adsorbed serum proteins, RGD and proteoglycan-binding peptides on the adhesion of mesenchymal stem cells to hydroxyapatite, Biomaterials 28 (2007) 383–392.

[84] V. Castelletto, R.M. Gouveia, C.J. Connon, I.W. Hamley, J. Seitsonen, A. Nykanen, et al. Alanine-rich amphiphilic peptide containing the RGD cell adhesion motif: a coating material for human fibroblast attachment and culture, Biomater Sci 2 (2014) 362–369.

[85] A. Horii, X. Wrang, F. Gelain, S. Zhang, Biological designer self-assembling peptide nanofiber scaffolds significantly enhance osteoblast proliferation, differentiation and 3-D migration, Plos One 2 (2007).

[86] R.G. Ellis-Behnke, D.K.C. Tay, et al. Nano neuro knitting peptide nanofiber scaffold for brain repair and axon regeneration with functional return of vision, Proc Natl Acad Sci 103 (2006) 5054–5059.

[87] B. Kao, K. Kadomatsu, H. Yoshiaki, Construction of synthetic dermis and skin based on a self-assembled peptide hydrogel scaffold, Tissue Eng 15 (2009) 2385–2396.

[88] D. Ho, M. Fitzgerald, C.A. Bartlett, B. Zdyrko, I.A. Luzinov, S.A. Dunlop, et al. The effects of concentration-dependent morphology of self-assembling RADA16 nanoscaffolds on mixed retinal cultures, Nanoscale 3 (2011).

[89] J. Liu, H. Song, L. Zhang, H. Xu, X. Zhao, Self-assembly-peptide hydrogels as tissue-engineering scaffolds for three-dimensional culture of chondrocytes *in vitro*, Macromolec Biosci 10 (2010) 1164–1170.

[90] Z. Yang, X. Zhao, A 3D model of ovarian cancer cell lines on peptide nanofiber scaffold to explore the cell-scaffold interaction and chemotherapeutic resistance of anticancer drugs, Int J Nanomed 6 (2011) 303–310.

[91] S. Zhang, T.C. Holmes, C.M. DiPersio, R.O. Hynes, X. Su, A. Rich, Self-complementary oligopeptide matrices support mammalian cell attachment, Biomaterials 16 (1995) 1385–1393.

[92] A. Mujeeb, A.F. Miller, A. Saiani, J.E. Gough, Self-assembled octapeptide scaffolds for *in vitro* chondrocyte culture, Acta Biomater 9 (2013) 4609–4617.

[93] L.A. Haines-Butterick, D.A. Salick, D.J. Pochan, J.P. Schneider, *In vitro* assessment of the pro-inflammatory potential of β-hairpin peptide hydrogels, Biomaterials 29 (2008) 4164–4169.

[94] Z. Luo, Y. Yue, Y. Zhang, X. Yuan, J. Gong, L. Wang, et al. Designer D-form self-assembling peptide nanofiber scaffolds for 3-dimensional cell cultures, Biomaterials 34 (2013) 4902–4913.

[95] S. Ustun, A. Tombuloglu, M. Kilinc, M.O. Guler, A.B. Tekinay, Growth and differentiation of prechondrogenic cells on bioactive self-assembled peptide nanofibers, Biomacromolecules 14 (2013) 17–26.

[96] M. Zhou, A.M. Smith, A.K. Das, N.W. Hodson, R.F. Collins, R.V. Ulijn, et al. Self-assembled peptide-based hydrogels as scaffolds for anchorage-dependent cells, Biomaterials 30 (2009) 2523–2530.

[97] V.N. Modepalli, A.L. Rodriguez, R. Li, S. Pavuluri, K.R. Nicholas, C.J. Barrow, et al. *In vitro* response to functionalized self-assembled peptide scaffolds for three-dimensional cell culture, Peptide Sci 102 (2014) 197–205.

[98] M.E. Davis, J.P.M. Motion, D.A. Narmoneva, T. Takahashi, D. Hakuno, R.D. Kamm, et al. Injectable self-assembling peptide nanofibers create intramyocardial microenvironments for endothelial cells, Circulation 111 (2005) 442–450.

[99] P.C.H. Hsieh, M.E. Davis, J. Gannon, C. MacGillivray, R.T. Lee, Controlled delivery of PDGF-BB for myocardial protection using injectable self-assembling peptide nanofibers, J Clin Investig 116 (2006) 237–248.

[100] Y.D. Lin, C.Y. Luo, Y.N. Hu, M.L. Yeh, Y.C. Hsueh, M.Y. Chang, et al. Instructive nano-fiber scaffolds with VEGF create a microenvironment for arteriogenesis and cardiac repair, Sci Transl Med 4 (2012).

[101] H. Hosseinkhani, M. Hosseinkhani, F. Tian, H. Kobayashi, Y. Tabata, Bone regeneration on a collagen sponge self-assembled peptide-amphiphile nanofiber hybrid scaffold, Tissue Eng 13 (2007) 11–19.

[102] R.E. Miller, A.J. Grodzinsky, E.J. Vanderploeg, C. Lee, D.J. Ferris, et al. Effect of self-assembling peptide, chondrogenic factors, and bone marrow-derived stromal cells on osteochondral repair, Osteoarthritis Cartilage 18 (2010) 1608–1619.

[103] C.J. Bell, L.M. Carrick, J. Katta, Z. Jin, E. Ingham, A. Aggeli, et al. Self-assembling peptides as injectable lubricants for osteoarthritis, J Biomed Mater Res A 78 (2006) 236–246.

[104] R. Mammadov, B. Mammadov, S. Toksoz, B. Aydin, R. Yagci, A.B. Tekinay, et al. Heparin mimetic peptide nanofibers promote angiogenesis, Biomacromolecules 12 (2011) 3508–3519.

[105] T.C. Holmes, S. de Lacalle, X. Su, A. Rich, Extensive neurite outgrowth and active synapse formation on self-assembling peptide scaffolds, Proc Natl Acad Sci 97 (2000) 6728–6733.

[106] V. Beachley, X. Wen, Polymer nanofibrous structures: fabrication, biofunctionalization, and cell interactions, Progr Polym Sci 35 (2010) 868–892.

CHAPTER 4

Self-Assembled Peptide Nanostructures for Regenerative Medicine and Biology

Ming Ni, Charlotte A.E. Hauser

Institute of Bioengineering and Nanotechnology, The Nanos, Singapore, Singapore

4.1 INTRODUCTION

Peptides are intriguing building blocks for a variety of applications in bionanotechnology. They are able to:

- Form nanostructures with distinct biological functions
- Fabricate biomimetic or bioinspired biomaterials at supramolecular level by *de novo* design
- Demonstrate many advantages over other building blocks.

Peptides are composed from a group of at least 20 naturally occurring amino acids, which comprise different functional side groups that constitute them chemically flexible and versatile. Peptides can be synthesized by solid phase and batch synthesis and easily scaled-up, depending on their size. In addition, synthetic peptide-based biomaterials are superior to many animal-derived biomaterials as they are generally nonimmunogenic and did not get into contact with any kind of pathogens. Short peptide epitopes can mediate biological recognition abilities. Furthermore, peptide-based nanostructured biomaterials offer favorable characteristics, since they are biodegradable, biocompatible, and nonimmunogenic because they are made of natural amino acids.

There are 23 proteinogenic amino acids. Twenty-one of them are encoded by the genetic code. These amino acids can be classified into different categories, for example, as hydrophilic and hydrophobic or as aliphatic, aromatic, polar, and charged amino acids. They can form linear peptides exerting secondary structures of α-helices, β-sheets, hair-pins, and β-turns. Branched and cyclic peptides can be synthesized as well. The organization of peptides into well-ordered nanostructures is facilitated by many noncovalent interactions, including electrostatic interactions,

Micro and Nanofabrication Using Self-Assembled Biological Nanostructures. DOI: 10.1016/B978-0-323-29642-7.00004-7

hydrogen bonds, hydrophobic interactions, van der Waals interactions, and π–π stacking interactions. Depending on their sequence, peptides can self-assemble into many forms of nanoscaled assemblies, including nanobelts [1], nano-doughnuts [1,2], nanofibers [1,3–5], nanomicelles [1], nanoribbons [1,6], nanorods [1], nanoropes [1], nanospheres [1,7], nanotubes [1,7–10], and nanovesicles [1,11]. In this chapter, we mainly focus on nanofibers and nanotubes, due to their relevance in biomedical applications. Their applications as tissue engineering scaffolds, drug delivery vehicles, and therapeutics will be illustrated. We also describe the nanofibers from a novel class of ultrashort peptides.

4.2 PEPTIDE BUILDING BLOCKS CONSTRUCTING NANOSTRUCTURES

4.2.1 Fabricating Nanofibers

4.2.1.1 Peptide Building Blocks with Alternating Hydrophobic and Hydrophilic Amino Acid Residues

Peptides with alternating hydrophobic and hydrophilic amino acid residues have the tendency to form β-sheet structure and further form nanofibers [12]. The repeating unit, (alanine-glutamic acid-alanine-glutamic acid-alanine-lysine-alanine-lysine)$_2$ (EAK16) was originally found in zoulin, a yeast protein. Circular dichroism spectra revealed that EAK16 peptides exhibit strong β-sheet secondary structures. Scanning electronic microscopy revealed that EAK16 peptides form interwoven fibrous network with fiber diameter of 10–20 nm.

4.2.1.2 Peptide Building Blocks with Alternating Polar and Nonpolar Amino Acid Residues

Peptides with alternating polar and nonpolar amino acid residues have the tendency to form either α-helix or β-sheet structures [13]. By carefully choosing the sequence periodicity of polar and nonpolar amino acids, we can design a peptide favoring particularly one of the α-helix or β-sheet secondary structures. Glutamic acid-leucine-lycine (ELK) is one of such examples. It forms strong β-sheet secondary structures and further forms nanofibers [14].

4.2.1.3 Ionic Self-Complementary Peptides

Another type of peptides that can form nanofibers is exemplified by the ionic self-complementary peptides. These peptides have been extensively studied by Shuguang Zhang's laboratory at MIT [15]. Ionic self-complementary peptide, RAD, contains regular repeating units of positively charged arginine (R) and negatively charged aspartic acid (D), and a spacing with alanine (A). When dissolved in aqueous solution, peptides form β-sheets structures with two distinct surfaces, one hydrophilic and the other hydrophobic, described as serving as pegs and holes similarly to Lego bricks that can be snapped together to assemble large constructs. The fiber diameter of self-complementary RAD peptides is approximately 10 nm. One of the RAD peptides, RADARADARADA (RADA16-I) was investigated for its reassembling process [16]. By applying mechanical agitation, such as sonication, RADA16-I nanofiber scaffolds are established in form of small fragments. However, these small fragments will quickly reassemble into a nanofiber scaffold with a different morphology. This reassociation/self-healing could be useful for tissue repair/regenerative medicine applications.

4.2.1.4 Collagen-Mimetic Peptides (CMP)

Type I collagen is the most abundant protein in our human bodies. It has triple-helix nanofiber structure. The building blocks for type I collagen are glycine-proline-X and glycine-X-hydroxyproline (G-P-X or G-X-O; X = any amino acid residues). Collagen has served as the target of biomimetic-designed structures for decades. Among all the designed molecules, Hartgerink's [17,18] multidomain peptides (MDPs) stand out as the best. Previously, Rele et al. [19] designed a 36 amino acid peptide with the sequence (proline-arginine-glycine)$_4$(proline-hydroxyproline-glycine)$_4$(glutamic acid-hydroxyproline-glycine)$_4$. Hartgerink [18] proposed to replace the arginine residues with lysine and the glutamate residues with aspartate and so the sequence became (proline-lysineglycine)$_4$(proline-hydroxyproline-glycine)$_4$(aspartic acid-hydroxyproline-glycine)$_4$. The charge pairing of lysine and aspartate formed an intimate salt-bridge hydrogen bond, serving as a sticky end, to stabilize the collagen-like triple helical structure. This peptide can self-assemble into nanofibers. The nanofibers will form hydrogels and the hydrogel can be degraded by collagenase at a similar rate to that of rat-tail collagen. In short, the designed peptide recapitulates the multi-hierarchical assembly of natural collagen.

Yu's laboratory designed a short CMP with the canonical proline-hydroxyproline-glycine (P-O-G) triplet repeats found in natural collagen [20]. This peptide exhibits strong binding affinity to both natural and denatured collagen. As such, fluorescent-labeled CMP [21] was used as an alternative to anticollagen I antibody to visualize collagens in fixed tissue sections. *In vivo* studies showed that CMP-based probes can detect abnormal bone growth activity in a mouse model of Marfan syndrome [22].

4.2.1.5 Coiled-Coil α-Helices

The α-helix is another valuable building block for constructing fibrous biomaterials [23,24]. Woolfson's laboratory [23] proposed a rational design of linear peptides that can form coiled-coil α-helices and further self-assemble into nanofibers. This is also known as a self-assembling fiber (SAF) system [23]. In such a system, two complementary leucine zipper (LZ) peptides are designed to coassemble to form sticky-end dimers. The LZ peptides contain heptad repeat sequence, often denoted as *abcdefg*. For example, isoleucine and leucine residues are positioned at *a* and *d,* best specifying dimeric LZ. These dimers assemble end-to-end and form long, noncovalently α-helical coil-coiled fibrils [23,24]. The diameters of the fibers are between 10–40 nm [24]. Recently, Woolfsen's laboratory formed caged structures from coiled-coil peptide modules [25]. These cages could potentially be used as delivery vehicles for bioactive molecules.

4.2.1.6 ABA Tri-Block Multidomain Peptides

In Hartgerink's laboratory, an ABA tri-block MDP was designed to form nanofibers [26]. The central "B" block contains alternating hydrophilic and hydrophobic amino acids. The peripheral "A" block is usually made of charged amino acid, such as lysine, to provide electrostatic interactions. This peptide adopts β-sheet conformation and self-assembles into nanofibers. The effects of replacing aliphatic amino acids with aromatic amino acids in the central domain of ABA peptides were investigated [17]. Choosing aromatic amino acids rendered antiparallel hydrogen bonding, whereas selecting aliphatic amino acids rendered parallel hydrogen bonding. This not only changed the nanofiber morphology, but also the hydrogel's rheological properties.

4.2.1.7 Peptide Building Blocks with Multidomains

Stupp's laboratory designed a subset of peptide amphiphiles (PAs) that form self-assembled nanofibers [27,28]. A representative structure of a

PA consists of three regions or structure domains: (1) of a hydrophobic tail, (2) of β-sheet forming amino acids, and (3) of one or more charged residues to improve solubility. PA nanofibers are typically 5–15 nm in width and 10 μm in length. Stupp's laboratory has changed the hydrophobic tail chain length and peptide amino acid compositions, finding that these changes did not alter a molecule's ability to self-assemble into nanofibers, but instead changed the fiber's morphology [29]. A bioactive domain can be added to the three-domain PA in order to render its bioactivity. Examples include a peptide sequence of arginine-glycine-aspartic acid (RGD) [28].

4.2.2 Fabrication Nanotubes

Peptide nanotubes can be constructed by the self-assembly of flat, ring-shaped peptide subunits composed of alternating D- and L-α-amino acid residues [30,31]. The subunits could be linear or cyclic peptides. The antibiotic gramicidin A is a linear peptide built of alternating D- and L-α-amino acids that fold in lipid membranes into tubular β-helixes [32]. Cyclic-D,L-peptides are other examples that can form nanotubes via macrocyclization. This was used to design artificial transmembrane ion channels. First reported by Ghadiri et al. [33], an octacyclic peptide assembled to an artificial transmembrane ion channel, showing that transport activity for K^+ and Na^+ was greater than 10^7 ions/s. The inner diameter of these tubular structures can be tuned by varying the size of peptide subunits [31]. For example, the diameter was varied from 7 to 13 Å by changing the number of amino acids in the cyclic peptide subunits from octacyclic peptides to 12-mer cyclic peptides [34,35]. Recently, Montenegro et al. [36] coupled single-walled carbon nanotubes (SWCNTs) to self-assembling cyclic peptide nanotubes (SCPN) to form SWCNT/SCPN hybrids. The formation of such hybrid nanostructures could provide synergistic properties derived from each individual and complementary nanostructures.

Coiled-coil helices are also able to form nanotubes. Recently, Woolfson's laboratory [37] designed coiled-coil hexamer (CC-Hex) structures that can self-assemble into helical bundles with a central channel of 6 Å in diameter. These molecular building blocks can be used for ion channels, nanoelectronic circuits and devices.

Linear surfactant-like hepta- and octapeptides can self-assemble into nanotubes or nanovesicles [11,15,38]. The design of the peptides mimics the physical properties of phospholipids. This class of peptides contains a hydrophilic head group consisting of one or two charged amino acids and a tail consisting of six hydrophobic amino acids. Examples include Ac-VVVVVVDD (V6D2), Ac-VVVVVVKK (V6K2), and Ac-AAAAAAK (A6K) [11,15,38]. The molecular assembly process of surfactant peptides mirrors that of lipid surfactants but at one order magnitude smaller in diameter.

Diphenylalanine (FF) has been suggested to be the core recognition motif of the β-amyloid polypeptide. The FF motif can self-assemble into two types of nanostructures: nanotubes [5] or nanospheres [7,39] depending on the preparation methods. Reches and Gazit [5] first reported the synthesis of FF nanotubes. The dipeptide building blocks self-assemble into nanotubes possibly by π–π stacking and β-sheet secondary structure formation. These nanotubes are chemically and thermally stable, having extraordinary mechanical strength. They have been used as a mold to cast silver nanowires [5]. Gazit's group [40] also achieved two-dimensional nanotube arrays (nanoforest) with either vertical or horizontal patterns. Via rapid evaporation, vertical patterns of FF nanotube arrays can be achieved. By first coating FF nanotubes with magnetic particles and then applying an external magnetic field, horizontal patterns of FF nanotube arrays can be achieved.

4.2.3 Ultrashort Peptides as Building Blocks to Construct Nanostructures

Recently, Hauser and her laboratory at the Institute of Bioengineering and Nanotechnology (IBN) in Singapore discovered a family of ultrashort linear peptides with 3–7 natural aliphatic amino acids [3,4]. These peptides can self-assemble to helical fibers within supramolecular structures [3,4]. The novel class of peptides shares a characteristic peptide sequence: it is amphiphilic, consisting of an aliphatic amino acid tail of decreasing hydrophobicity and a hydrophilic head (Figure 4.1). These peptides self-assemble most likely via parallel–antiparallel α-helical pair formation and subsequent stacking into fibers that condense to β-turn fibrils [3], as illustrated in Figure 4.1. Fibrils further aggregate and form nanofibrous scaffolds – macroscopic hydrogels.

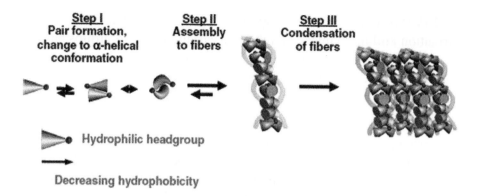

Fig. 4.1. Hypothesis of self-assembly from peptide monomers to supramolecular networks of condensed fibers. Self-assembly is initiated with antiparallel pairing of two peptide monomers by changing to α-helical conformation. Subsequently, peptide pairs assemble to fibers and nanostructures and condense to fibrils resulting in hydrogel formation. Figure reproduced from Ref. [3]. Copyright 2011 National Academy of Sciences, USA.

Hydrogels from designed peptides have many advantages over other biomaterials: (1) their water content is high, entrapping over 99% of water, (2) they are designed at a molecular level, they are fully synthetic and therefore nonimmunogenic, unlike animal-derived biomaterials that may in addition carry dangerous pathogens, (3) they contain both amine and carboxylic acid groups that allow functionalization and the addition of bioactive moieties, (4) they are soft matters, suitable to replace or repair tissues, and (5) they can be used as scaffolding materials to encapsulate cells, genes, and drugs. However, most peptide hydrogels are very expensive when considering the costs of their synthesis due to their extensive length (with amino acid residues ≥16). By contrast, the ultrashort peptide hydrogels developed by Hauser et al. are made only of 3–7 amino acids, which make the synthesis easy and cost-effective. Moreover, no enzymes, physical or chemical cross-linkers are needed to form hydrogels. The ultrashort peptide hydrogels have demonstrated high mechanical stiffness, thermal stability, and biocompatibility [4], making them attractive for various biotechnological and industrial applications.

In Hauser's laboratory, more than 100 amphiphilic ultrashort peptides were rationally designed and evaluated [4]. Among them, around 50 hydrogel-forming peptides were investigated in detail. The ability of hydrogel formation is in an order of hexamer > heptamer > pentamer > tetramer > trimer [3]. The aliphatic trimers Ac-IVD and Ac-IID are the smallest amphiphilic hydrogel-forming peptides [3]. Hexamer

Ac-LIVAGD (LD$_6$) was the best-performing hydrogel for ease of hydrogel formation and mechanical strength [4]. Varying the head groups has an effect on the gelation properties of the resulting hydrogels in the order of acidic (D and E) >neutral (S and T) >basic (K) polar, non-aromatic amino acids [3].

Each ultrashort peptide has a critical gelation concentration [4]. With increasing concentration, the ultrashort peptide hydrogels changed from a clear to a translucent and then to an opaque appearance. Ultrashort peptides form nanofibers and nanostructures revealed by field-emission scanning electron microscopy (FESEM). The diameters of these fibers were around 50–60 nm, similar to natural collagen. Under different concentration, peptides showed different morphologies. For example, LD$_6$ peptide hydrogels showed web-like nanostructure at 1 mg/mL, whereas the morphology was changed to honeycomb nanostructure at 20 mg/mL. A rheometer measured the viscoelastic properties of peptide hydrogels. LD$_6$ hydrogels demonstrated 10-fold higher mechanical strength than collagen [4].

Ultrashort peptide hydrogels also showed cytocompatibility to many cell types, including human mesenchymal stem cells, human embryonic stem cells, human primary renal tubular cells, human umbilical vein endothelial cells (ECs), rat hepatic stellate cells, rabbit fibroblasts, and rabbit retinal epithelial cells [4]. In addition, peptide hydrogels were non-hemolytic.

The effects of various metal salts on the gelation properties of the peptides were investigated [41]. It was found that the majority of metal salts did not interrupt the hydrogelation process. However, trivalent salts, such as FeCl$_3$, caused the generation of precipitates. These results indicated that peptide hydrogels can remain stable in the presence of common physiological metal salts, which make them suitable for many biomedical applications.

4.3 CELL ADHESION

Hydrogels have been extensively studied as scaffold materials for tissue engineering and regenerative medicine [42,43]. Most hydrogels are biocompatible and nonadhesive to cells [42]. Therefore, it is important

to introduce cell adhesive cues to hydrogels to guide cell attachment and functions. As mentioned previously, peptide hydrogels are superior to common biomaterials as they can be functionalized and designed at molecular level. They also share similar nanofibrous structures as extracellular matrix (ECM) proteins, such as type I collagen, elastin, and others. Cell adhesion to the ECM occurs via binding of integrin receptors to specific epitopes on the surface of ECM proteins, such as fibronectin, vitronectin, and laminin, leading to formation of focal adhesions and related contacts. Amino acid sequences of arginine-glycine-aspartic acid (RGD) and tyrosin-isoleucine-glycine-serine-arginine (YIGSR) are commonly used as they are derived from fibronectin and laminin [43].

In Hubbell's laboratory [44], peptide-coated glass was used as model surfaces to study the effect of RGD ligand density on fibroblasts spreading. They found that the complete spreading of fibroblasts required 10 fmol/cm^2 of RGD. In Stupp's laboratory [45], both branched and linear PA nanofiber with RGD epitopes were used to study the architectural effect on cell adhesion, spreading, and migration. They found that branched PA gave better results. Zhou et al. [46] mixed two types of building blocks, FF and RGD, as one of the simplest self-assembling moieties to form hydrogels. The RGD ligand density on the fiber surfaces can be tuned accordingly. Therefore, this may offer an economical model scaffold to three-dimensional (3D)-culture many anchorage-dependent cells for *in vitro* tissue regeneration. Luo et al. [47] used chiral D-form self-assembling peptides as scaffold for 3D cell culture. The authors believe that D-form peptides may have many advantages over L-form peptides, including resistance to protease degradation. These peptides were used to stop bleeding quickly and support cell growth.

Different surface modification approaches have been applied to enhance cell adhesiveness to biomaterials [43,48]. For example, by applying (1,1-carbonyldiimidazole) (CDI) chemistry, one can conjugate peptide or proteins to a surface [48,49].

In Hauser's laboratory [50], cysteine was added to modify the ultrashort peptide Ac-LIVAGK (LK$_6$). Introducing the disulfide bond further stabilized peptide hydrogel formation. The presence of chemical cross-links increased the gels' ability to maintain their shape when

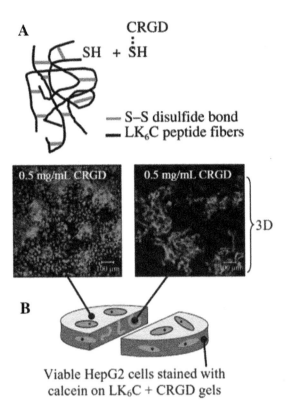

Fig. 4.2. *Cross-linking and biofunctionalization strategy. (A) Scheme to introduce chemical cross-links among LK_6C peptide fibers and their conjugation with CRGD, the integrin-recognition motif, via cysteine-mediated disulfide bonds. (B) HepG2 cells were viable, as evident from the positive calcein signals, after 4 days of culture on the RGD-functionalized, cross-linked and purified LK_6C gels. 3D distribution was also achieved, as confirmed by cross-sectional slices, which revealed multilayered cell growth.* Adapted with permission from Ref. [50]. Copyright 2013 Wiley-VCH.

soaked in media during prolonged culture (Figure 4.2). Additionally, the cell adhesion motif, CRGD, was conveniently tethered onto the peptide backbone via disulfide bridges. Such RGD-functionalized cross-linked gels enhanced cell proliferation, further supporting the cells' infiltration into the porous gel to grow in a 3D environment (Figure 4.2).

4.4 TISSUE ENGINEERING

4.4.1 Orthopedic Applications

Bone is a composite material made of an organic matrix and inorganic calcium phosphate. The major component of the bone matrix is collagen type I. As discussed previously, there are different approaches to mimic the synthesis of natural collagen using synthetic peptides [17–20]. Mineralization of biocompatible peptide hydrogels is of interest for orthopedic applications because these composites mimic bone structure. Several approaches have been investigated. Here are a few listed:

- Song et al. [51] introduced carboxyl groups to the surface of poly(2-hydroxyethyl methacrylate) (pHEMA) hydrogels and found that it can promote calcium phosphate nucleation and growth on pHEMA surfaces. A mineral layer of a few micrometers thickness was formed.
- Hartgerink et al. [52] designed peptide amphiphilic sequences containing phosphorylated serine. Hydroxyapatite (one type of calcium phosphate) is able to grow along the axis of peptide nanofiber (epitaxial crystallization), which is similar to that of natural bones having hydroxyapatite crystals aligned in parallel with collagen fibrils. However, in osteogenic medium supplemented with calcium, spherical hydroxyapatite nanoparticles approximately 100 nm in diameter were formed [53]. This is possibly due to a rapid amorphous calcium phosphate precipitation. It indicates that the conditions for mineralization are very important. Although artificial peptides designed with calcium-binding motif, such as phosphorylated serine, can mimic natural bone matrix to certain degrees under a well-defined buffer system, these peptides cannot replace the natural bone matrix when changing to a more complex buffer system.
- Mata et al. [54] filled the rat femoral critical-size defect (a 5-mm gap) with preassembled phosphorylated serine containing PA nanofiber hydrogels and analyzed bone formation with microcomputed tomography and histology. Within 4 weeks, they found that bone formation was comparable to that observed in animals treated with a clinically used allogenic bone matrix. Therefore, these PA hydrogels are interesting candidates to replace autografts or allografts.
- Bone morphogenetic protein-2 (BMP-2) is one of the most important growth factors involved in bone regeneration. Lin et al. [55] showed that a BMP-2-analog peptide, P24, can enhance osteoblasts (bone-forming cells) differentiation *in vitro*. When P24

was incorporated in a biodegradable polymer, sustained release of P24 promoted bone formation *in vivo*.

- Lutolf et al. [56] engineered poly(ethylene glycol)-based hydrogel as cell-in-growth matrices for bone regeneration. Their hydrogel network contains an integrin binding RGD ligand to foster cell adhesion and a matrix metalloproteinase (MMP)-cleavable site that makes the hydrogel degradable. BMP-2 was encapsulated in gels to enhance the bone regeneration. This multifunctional hydrogel healed critically sized defects in rats.
- Horii et al. [57] functionalized RADA-16 with three short peptide motifs: (i) osteogenic growth motif, ALK, is responsible for bone tissue formation; (ii) osteopontin motif, DGR, is involved in bone mineralization; and (iii) RGD-based motif, PGR, is involved in binding cells. When short segments of ALK, DGR, and PGR were chemically coupled via solid-phase peptide synthesis to the RADA16 peptide, the resultant peptide-functionalized hydrogels not only promoted proliferation of mouse preosteoblasts, but also their osteogenic differentiation. These observations highlight the importance of osteoblast attachment to the scaffold in both cellular processes and illustrate how these scaffolds can be useful in bone regeneration and bone tissue engineering.
- Degenerative disc disease is the predominant cause of disability in the adult population, affecting 85% of the population by age 50. There is a huge unmet clinical need for disc prosthesis that can inhibit or repair early-stage disc damage. Mishra et al. [4] proposed using ultrashort peptide hydrogels to repair early-stage disc damage. These hydrogels are cyto-compatible toward porcine nucleus pulposus (pNP) cells and human mesenchymal stem cells, suggesting that these new materials can be suitable for tissue regeneration in orthopedic and plastic surgery.

4.4.2 Wound Healing

In the United States, approximately 450,000 individuals are getting hospitalized each year because of burns and roughly 4,500 people from them die [58]. Several tissue-engineered skin grafts have been developed [59], including TransCyte and Oasis Wound Matrix. Loo et al. [60] used ultrashort peptide hydrogels as a primary dressing to treat partial thickness burns in a rat model. Compared to a commercial dressing, Mepitel®, ultrashort peptide hydrogel-based dressing showed faster wound closure (Figure 4.3). In addition, ultrashort peptides have shown to exert long shelf-life stability at room temperature. Therefore, "just-add-water" formulations can be developed. The hydrogels can be reconstituted by adding a fixed volume of sterile water to lyophilized

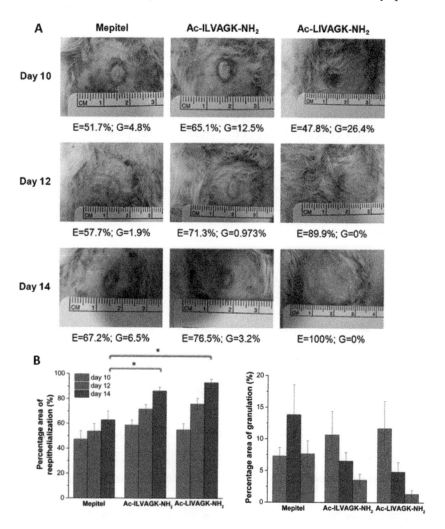

Fig. 4.3. (A) Both Ac-ILVAGK-NH₂ and Ac-LIVAGK-NH₂ hydrogels accelerated the regeneration of new epidermal tissue, as denoted by the area of reepithelialization, E. The granulation, G, also decreased over time. All the images in this figure show the progression of wound healing for the same animal (H24). The burn injury treated with Ac-LIVAGK-NH₂ hydrogel completely regenerated its epidermis by day 14. (B) Quantitative evaluation of wound healing using digital planimetry revealed that the peptide hydrogels stimulated reepithelialization and reduced granulation, compared to Mepitel. The error bars denote the standard error of the mean (n = 6). Figure reproduced from Ref. [60]. Copyright 2014 Elsevier.

peptide powder at the point of application. This formulation design greatly reduces transportation costs and potentially revolutionizes emergency medicine, providing a convenient, easy-to-use first-line treatment for partial thickness burn injuries.

Another important issue related to wound healing is to prevent infections. Reithofer et al. [61] developed antibacterial hydrogels that show sustained release of newly synthesized silver nanoparticles. Silver nanoparticles were synthesized by simple UV irradiation. Ultrashort peptide hydrogels were used as matrices to control the size and prevent aggregation of silver nanoparticles (Figure 4.4A, B). Silver nanoparticles were released from ultrashort peptide hydrogels in a sustained manner over 14 days (Figure 4.4C). The released silver particles can inhibit both gram-positive and gram-negative bacterial growth (Figure 4.4D). Moreover, no cytotoxicity was observed when culturing human fibroblasts on these silver-releasing hydrogels. Based on these results, the authors have recommended this composite biomaterial to be used for wound healing applications.

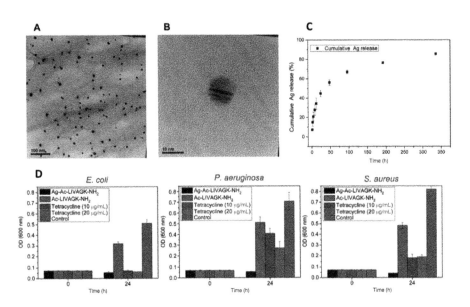

Fig. 4.4. Antibacterial hydrogels containing silver nanoparticles. (A, B) TEM images of the silver nanoparticles distribution in Ag-Ac-LK$_6$-NH$_2$ hydrogels formed in water using 10 mM AgNO$_3$ solution; (C) cumulative release of silver from 100 µL of Ag-Ac-LK$_6$-NH$_2$ hydrogels as a function of time; (D) graphical representation of the OD measurements in Escherichia coli, Pseudomonas aeruginosa, and Staphylococcus aureus at t = 0 and t = 24 h. Figure reproduced from Ref. [61]. Copyright 2014 Elsevier.

4.4.3 Cartilage Tissue Engineering

Approximately one million people require treatment of cartilage defects every year in the United States [62]. Tissue engineering approaches offer new possibilities for functional and structural restoration of damaged or lost cartilage tissues. Shah et al. [63] designed a PA molecule with the peptide sequence HSNGLPL that displays a high-binding affinity to transforming growth factor β-1 (TGFβ-1). This PA molecule was coassembled with a nonbioactive PA molecule to form nanofibers. The resulting nanofiber scaffold was used for cartilage regeneration. Human mesenchymal stem cells were cultured within the PA hydrogels *in vitro*. Cells survived and underwent chondrogenic differentiation. In a full thickness chondral defect model, the PA hydrogels promoted the regeneration of articular cartilage. Kisiday et al. [64] developed a self-assembling peptide KLD-12 based hydrogel as a 3D scaffold for encapsulating chondrocytes. In an *in vitro* experiment, chondrocytes seeded within the peptide hydrogel retained their morphology and secreted proteoglycans and collagen type II, thereby indicating a stable chondrocyte phenotype.

4.4.4 Liver Tissue Engineering

Liver disease is one of the leading causes of mortality and results in the United States in over 26,000 deaths annually [65]. Due to limited organ donors, tissue engineering provides an alternative to replacing tissue or organ functions lost to disease. Semino et al. [66] reported the differentiation properties of a putative rat liver progenitor cell line cultured in a self-assembling peptide scaffold, RADA16-I. Peptide scaffold supported hepatic differentiation. Cells showed several characteristics of mature hepatocyte behavior, including upregulation of albumin and expression of inducible CYP1A1, CYP1A2, and CYP2E1 cytochrome p450 enzymes that can produce complex metabolic products. Thus, these cells can be used for testing human toxicity and drug efficacy.

Mehta et al. [67] studied the effects of the mode of epidermal growth factor (EGF) presentation (soluble versus tethered) on primary rat hepatocyte using RADA peptide hydrogels with adhesive ligands. WQPPRARITGY and RGD motifs were selected to enhance cell adhesiveness. In the presence of soluble EGF, hepatocytes cultured on RADA peptide hydrogel could aggregate, spread, and maintain metabolic function but failed to induce DNA synthesis. On the contrary, tethered EGF could induce DNA synthesis. Phenotypic differences between soluble and tethered EGF stimulation of cells on peptide gels correlated with differences in expression and phosphorylation of the EGF receptor and its heterodimerization partner ErbB2. This enabled the authors to study the synergistic effect of adhesion ligand and growth factor receptors.

4.4.5 Pancreatic Islet Transplantation

Diabetes is a leading cause of mortality [68] with over 25 million people in the United States having the disease [68]. Pancreatic islet transplantation offers a suitable substitute for daily administration of insulin, and thus can lower blood glucose levels to treat diabetes. Khan et al. [69] used glucagon-like peptide 1 (GLP-1)-mimetic PA to self-assemble into nanofibers and form hydrogels. Pancreatic β cells were encapsulated in 3D gels. The GLP-1-mimetic PA nanofibers stimulated insulin secretion and promoted β cells' proliferation. The PA hydrogel scaffold holds the promise to support the survival, proliferation, and function of transplanted β-cells during the post-transplant period. Lim et al. [70] created an ECM mimic material decorated with bioinductive cues for β-cells by using self-assembled PAs containing four selected ECM-derived cell adhesive ligands: (1) RGD found in collagen IV and laminin-1, (2) isoleucine-lysine-leucine-leucine-isoleucine (IKLLI), (3) isoleucine-lysine-valine-alanine-valine (IKVAV) in the α1 chain of laminin-1, and (4) YIGSR found in the β1 chain of laminin-1. The functionalized PA nanomatrices lead MIN6 β cells to enhance insulin secretion function from glucose stimulation. These findings might improve survival and function of transplanted islets used in treating type 1 diabetes.

4.4.6 Vascular Tissue Engineering

Cardiovascular diseases are considered as the top killers in our modern society. Every year, more than a half-million coronary bypass grafts are performed in the United States [71]. Due to the autograft shortage, synthetic vascular grafts, Dacron and polytetrafluoroethylene (PTFE) are used. However, synthetic grafts usually fail due to thrombosis and restenosis. Thus, developing tissue-engineered blood vessels is imperative. The success of tissue-engineered grafts largely depends on its ability to replicate the microenvironment of native tissue. ECs and smooth muscle cells (SMCs) form the cellular components of a blood vessel. The critical component of the tissue-engineered vessels is a functional endothelium, a collagenous network for mechanical strength and an elastin network to recreate the mechanical flexibility of native blood vessels. To this end, peptide-based nanomatrices have been developed to mimic the natural ECM and form a confluent EC layer. The EC layer provided the best antithrombogenic surface. These nanomatrices were self-assembled by three PAs, one containing a laminin-derived peptide motif YIGSR, the second containing an elastin-derived peptide motif valine-alanine-proline-glycine (VAPG) ligand, and a third did not contain any cell adhesive ligands, C_{16}-GTAGLIGQS (PA-S) [72,73]. YIGSR was chosen to enhance EC adhesion, spreading, and proliferation, which are essential for developing a functional endothelium. VAPG has also been shown to assist SMC adhesion and spreading, which plays an important role in developing a SMC layer, imparting mechanical integrity to the graft and aiding in recreating the vascular trilayer. Furthermore, these PAs did not support platelet adhesion and therefore may not cause thrombosis. These PAs could facilitate development of novel vascular grafts, heart valves, and potentially novel cell-based therapies for cardiovascular diseases.

McClendon and Stupp [74] constructed an arterial tubular scaffold by bundling PA nanofibers. Human coronary artery SMCs were encapsulated in the tubular scaffold and gained circumferential alignment spontaneously. The advantage of using PA molecules is the possibility of adding bioactive signals at peptides' termini, thus allowing biological signaling to further control SMC behavior. A number of bioactive PAs have been designed for regenerative medicine, including some with the specific ability to bind growth factors [63,75], promote differentiation [52], and accelerate angiogenesis [76].

4.4.7 Neuronal Tissue Engineering
4.4.7.1 Peripheral Nerve Injury
Peripheral nerves can regenerate after injury. However, when the gap between injured nerves is too big, synthetic nerve conduits are needed [62]. Synthetic nerve conduits made of natural polymers (laminin, collagen, chondroitin sulfate) or synthetic polymers were shown to enhance nerve regeneration [62]. Peptide-based biomaterials were also investigated in this regard. In one study, Sur et al. [77] mixed collagen type I and

PA molecules containing IKVAV and YIGSR epitopes. Varying epitope density modulated neuron survival and maturation. This provides a versatile test bed to study the ECM contribution in neuron development and the design of optimal neuronal scaffold biomaterials.

Holmes et al. [78] used RADA-based peptide hydrogels as scaffolds to culture primary rat neurons. They found (i) the scaffolds can support neuronal cell attachment and differentiation as well as extensive neurite outgrowth, (ii) the scaffolds are permissive substrates for functional synapse formation between attached neurons. This suggests that these scaffolds could be useful for neuronal tissue engineering applications. Koutsopoulos and Zhang [79] designed another type of functionalized self-assembling RADA-based peptide hydrogel as a neural tissue-engineering scaffold. It contains the peptide sequences of SKPPGTSS, PFSSTKT, and RDG motifs. Both SKPPGTSS and PFSSTKT motifs were selected for the enhancement of neuronal cell adhesion and differentiation. These designed 3D-engineered tissue culturing systems have a potential use for neuronal tissue regeneration and fundamental biological studies of neural cells in a biomimetic environment.

4.4.7.2 Spinal Cord Injury

To repair spinal cord injuries is a much more challenging task than repairing peripheral nerve injury. This is caused, because axons in the central nerve system don't regenerate [80]. Therefore, new strategies were developed aiming at coaxing disconnected axons to regrow across spinal cord lesions [81]. Silva et al. [82] designed a PA molecule containing IKVAV, a laminin-derived motif. PA formed nanofibers and macroscopic hydrogels by self-assembly. Neural progenitor cells were encapsulated *in vitro* within PA hydrogels. These hydrogel materials induced very rapid differentiation of cells into neurons, while discouraging the development of astrocytes. This was justified because PA nanofibers in the hydrogel network contain a much higher density of laminin motifs compared to that of the adsorbed laminin protein. PA system has an estimated 7.1×10^{14} IKVAV epitopes/cm^2. By contrast, adsorbed laminin has an estimated 7.5×10^{11} IKVAV epitopes/cm^2. Thus, the IKVAV nanofibers of the network can amplify the epitope density relative to a laminin monolayer by roughly a factor of 10^3. These PA solutions were injected *in vivo* into rats to treat spinal cord injuries.

Iwasaki et al. [83] injected neural stem/progenitor cells (NPCs) together with QL6 peptide into the lesion epicenter 2 weeks after bilateral clip compression-induced cervical spinal cord injury (SCI) in rats. The combination of QL6 and NPCs promoted forelimb neurobehavioral recovery and was associated with significant improvement in forelimb print area and stride length. Spinal cord tissue was significantly preserved at 12 weeks after injury, which contributed to the significant recovery of forelimb neural function. Thus, the use of self-assembling peptides is a promising treatment for SCI from the viewpoint of engineering a relevant scaffold and providing neuroprotective effects. The QL6 is characterized by periodic repeats of alternating ionic hydrophilic and hydrophobic amino acids (glutamine and leucine) and can self-assemble into hydrogels.

4.4.7.3 Models to Study the Development of Hippocampal Neurons

The mechanical properties of the ECM are known to influence neuronal differentiation and maturation. Sur et al. [84] used self-assembled peptide nanofibers to mimic ECM. Nanofiber rigidity was tailored by supramolecular interactions to investigate the relationship between matrix stiffness and morphological development of hippocampal neurons. Tuning matrix stiffness can enhance the therapeutic potential of these materials in applications for regenerating the nervous system.

4.4.8 Cell and Organ Printing

Soft lithography was used to pattern and align PAs nanofibers [85]. Mata et al. [86] incorporated polymerizable acetylene groups in the hydrophobic segment of PAs. With this, they could micropattern nanofiber gels. PAs containing the cell adhesive epitope arginine-glycine-aspartic acid-serine (RGDS) were allowed to self-assemble within microfabricated molds to create networks of either randomly oriented or aligned 30 nm diameter nanofiber bundles that were shaped into topographical patterns containing holes, posts, or channels up to 8 mm in height and down to 5 mm in lateral dimensions. When topographical patterns that were contained in nanofibers aligned through flow before gelation, the majority of human mesenchymal stem cells aligned in the direction of the nanofibers. Even in the presence of microtextures, more than a third part of cells maintained this alignment when encountering perpendicular channel microtextures. Interestingly, in topographical patterns with randomly oriented nanofibers, osteoblastic differentiation of human mesenchymal stem cells was enhanced on microtextures compared to all other surfaces.

Inkjet printing, present in households and most offices, is a common method for transferring digital data to paper or transparencies. The aromatic dipeptides, FF, can form both spherical and tubular structures. These structures were used as bio-ink to be efficiently patterned on surfaces via a desktop inkjet printer [87]. As discussed above, a variety of cells can be encapsulated into the peptide hydrogel. Therefore, in principle, cell-containing peptide hydrogels can be patterned on the surface either by soft lithography or inject printing.

4.5 PEPTIDE HYDROGELS AS VEHICLES FOR CONTROLLED DRUG DELIVERY

It is an important goal when using drug delivery systems to control the drug's duration of action and level in the human body [88]. Prevalent mechanisms for the drug delivery systems are dissolution of the vehicles and diffusion. Peptide-based hydrogels are excellent controlled release materials because they are generally biodegradable and biocompatible.

Matson et al. [27] used PA nanofibrous hydrogels as biodegradable drug delivery vehicles. A dye, 6-propionyl-2-dimethylaminonaphthalene (Prodan), was used as a model drug. The release of Prodan from hydrogels was measured. Near zero-order release kinetics were observed. Relative release rates correlated directly with fluorophore mobility, which varied inversely with packing density, degree of order in the hydrophobic PA core and the peptide's β-sheet character. Koutsopoulos et al. [89] measured protein release from RADA-based hydrogels. This RADA system would allow the release of diffusing molecules in a sustainable highly efficient manner. The protein release rate can be tuned by changing hydrogel nanofiber density to control the release kinetics. Webber et al. [90] conjugated an anti-inflammatory drug, dexamethasone (Dex), to a modular PA molecule via a hydrazone linkage. This molecule self-assembled in water into long supramolecular nanofibers when mixed with a similar PA without the drug conjugate. This Dex–PA nanofiber gel showed anti-inflammatory activity both *in vitro* and *in vivo*. Such drug-conjugated PA could be generalized to many targets in regenerative medicine and also adapted to other peptide gelator systems as it is highly amenable to solid-phase peptide synthesis and can conceivably be used to conjugate any ketone- or aldehyde-containing drug to virtually any peptide-based materials.

Ultrashort peptides can be used for drug delivery. Reithofer et al. [91] investigated click chemistry to attach the anticancer drug oxaliplatin to the ultrashort peptide LK_6 (Figure 4.5). When mixing the oxaliplatin-conjugated peptide with pristine peptide, hydrogels formed that held up to 40% drug loading. In a mouse model, the oxaliplatin-conjugated peptide showed tumor growth inhibition and significantly lower toxicity compared to the free drug (Figure 4.6). Click chemistry provides a versatile approach to functionalize peptide-based materials with a variety of different bioactive compounds.

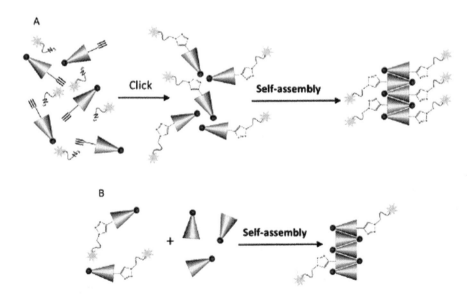

Fig. 4.5. Schematic drawing of (A) peptide functionalization with a bioactive cue using click chemistry. The triangle represents the hydrophobic tail of the peptide showing decreasing lipophilicity from N- to C-terminus and the dark gray (red in the web version) dot represents the polar head group at the C-terminus; (B) assembly of the parent peptide together with the functionalized ultrashort peptide, forming a hybrid system. Figure reproduced from Ref. [91]. Copyright 2013 Royal Society of Chemistry.

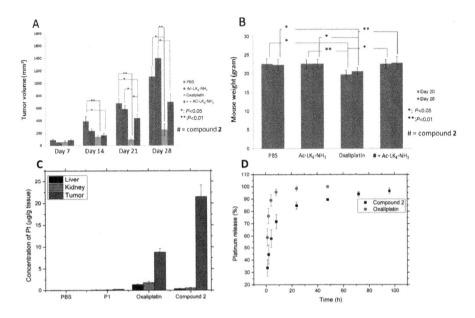

Fig. 4.6. Effects of oxaliplatin-derived peptide hybrid hydrogel in 4T1 tumor-bearing mice. (A) Tumor growth inhibition. (B) Reduced toxicity. (C) Biodistribution profiles of the injected compounds in the liver, kidney, and tumor of the treated animals. (D) In vitro drug release profile of the oxaliplatin-peptide conjugate and the free drug. Figure reproduced from Ref. [91]. Copyright 2013 Royal Society of Chemistry.

4.6 PEPTIDE THERAPEUTICS

Amyloid fibrils formation relates to many diseases such as Alzheimer's, Parkinson, and type 2 diabetes. Therefore, understanding amyloid formation at molecular level is essential. Recently, Hauser et al. [3] reported a class of rationally designed ultrashort peptides that self-assemble in water into amyloid-β type fibers. Lakshmanan et al. [92] compared the self-assembly process of these peptides with naturally occurring amyloidogenic core sequences from three amyloid-associated diseases, namely Alzheimer's, type 2 diabetes, and thyroid medullary carcinoma. The self-assembly process was monitored using different experimental assays such as circular dichroism, electron microscopy, X-ray diffraction, rheology, and molecular dynamics simulations. The designed aliphatic peptides formed α-helical intermediates, similar to several natural sequences. By contrast, the aromatic peptide FF, which was proposed to be a dominant part of the core of amyloid-β, showed distinctly different behavior compared to all other examined sequences. The diphenyl-alanine-containing sequence formed β-sheet aggregates without going through the α-helical intermediate step, giving a unique fiber diffraction pattern and simulation structure. In conclusion, aromatic interactions may not be as crucial in amyloid formation as thought. This study provides vital insight into the nature of early intermediates, which will facilitate searching for toxic species and potent hotspots of amyloid aggregation. It also provides the basis for developing therapeutic drugs that control and inhibit amyloid formation.

4.7 CONCLUSION

Peptides' unique properties offer great potential for a wide range of profound medical advances. More specifically, peptides can self-assemble into many nanostructures, including nanofibers and nanotubes. A novel class of ultrashort peptides that contain 3–7 natural aliphatic amino acids with an innate tendency to self-assemble can form hydrogels that non-toxic, highly thermal stable, and demonstrate high mechanical stiffness. These assemblies can be used as tissue engineering scaffolds, drug delivery vehicles, and therapeutics.

ACKNOWLEDGMENT

This work was funded by the Institute of Bioengineering and Nanotechnology (Biomedical Research Council, Agency for Science, Technology and Research, Singapore).

REFERENCES

[1] X. Zhao, F. Pan, H. Xu, C. Hauser, S. Zhang, J.R. Lu, Molecular self-assembly and applications of designer peptide amphiphiles, Chem Soc Rev 39 (2010) 3480–3498.

[2] R. Djalali, J. Samson, H. Matsui, Doughnut-shaped peptide nano-assemblies and their applications as nanoreactors, J Am Chem Soc 126 (2004) 7935–7939.

[3] C.A.E. Hauser, R. Deng, A. Mishra, Y. Loo, U. Khoe, F. Zhuang, et al., Natural tri- to hexapeptides self-assemble in water to amyloid beta-type fiber aggregates by unexpected alpha-helical intermediate structures, Proc Natl Acad Sci USA 108 (2011) 1361–1366.

[4] A. Mishra, Y. Loo, R. Deng, Y.J. Chuah, H.T. Hee, J.Y. Ying, C.A.E. Hauser, Ultrasmall natural peptides self-assemble to strong temperature-resistant helical fibers in scaffolds suitable for tissue engineering, Nano Today 6 (2011) 232–239.

[5] M. Reches, E. Gazit, Casting metal nanowires within discrete self-assembled peptide nanotubes, Science 300 (2003) 625–627.

[6] D.M. Marini, W. Hwang, D.A. Lauffenburger, S. Zhang, R.D. Kamm, Left-handed helical ribbon intermediates in the self-assembly of a β-sheet peptide, Nano Lett 2 (2002) 295–299.

[7] J. Castillo-Leon, R. Rodriguez-Trujillo, S. Gauthier, A.C.O. Jensen, W.E. Svendsen, Micro-"factory" for self-assembled peptide nanostructures, Microelectr Eng 88 (2011) 1685–1688.

[8] T. Aida, E.W. Meijer, S.I. Stupp, Functional supramolecular polymers, Science 335 (2012) 813–817.

[9] K.L. Niece, C. Czeisler, V. Sahni, V. Tysseling-Mattiace, E.T. Pashuck, J.A. Kessler, et al., Modification of gelation kinetics in bioactive peptide amphiphiles, Biomaterials 29 (2008) 4501–4509.

[10] Y. Ruff, T. Moyer, C.J. Newcomb, B. Demele, S.I. Stupp, Precision templating with DNA of a virus-like particle with peptide nanostructures, J Am Chem Soc 135 (2013) 6211–6219.

[11] S. Vauthey, S. Santoso, H. Gong, N. Watson, S. Zhang, Molecular self-assembly of surfactant-like peptides to form nanotubes and nanovesicles, Proc Natl Acad Sci USA 99 (2002) 5355–5360.

[12] S.G. Zhang, T. Holmes, C. Lockshin, A. Rich, Spontaneous assembly of a self-complementary oligopeptide to form a stable macroscopic membrane, Proc Natl Acad Sci USA 90 (1993) 3334–3338.

[13] H. Xiong, B.L. Buckwalter, H.-M. Shieh, M.H. Hecht, Periodicity of polar and nonpolar amino acid is the major derterminant of secondary structure in slef-assembling oligometric peptides, Proc Natl Acad Sci USA 92 (1995) 6349–6353.

[14] T.C. Holmes, Novel peptide-based biomaterial scaffolds for tissue engineering, Trends Biotechnol 20 (2002) 16–21.

[15] S.G. Zhang, Fabrication of novel biomaterials through molecular self-assembly, Nat Biotechnol 21 (2003) 1171–1178.

[16] H. Yokoi, T. Kinoshita, S.G. Zhang, Dynamic reassembly of peptide RADA16 nanofiber scaffold, Proc Natl Acad Sci USA 102 (2005) 8414–8419.

[17] E.L. Bakota, O. Sensoy, B. Ozgur, M. Sayar, J.D. Hartgerink, Self-assembling multidomain peptide fibers with aromatic cores, Biomacromolecules 14 (2013) 1370–1378.

[18] L.E.R. O'Leary, J.A. Fallas, E.L. Bakota, M.K. Kang, J.D. Hartgerink, Multi-hierarchical self-assembly of a collagen mimetic peptide from triple helix to nanofibre and hydrogel, Nat Chem 3 (2011) 821–828.

[19] S. Rele, Y. Song, R.P. Apkarian, Z. Qu, V.P. Conticello, E.L. Chaikof, D-periodic collagen-mimetic microfibers, J Am Chem Soc 129 (2007) 14780–14787.

[20] A.Y. Wang, X. Mo, C.S. Chen, S.M. Yu, Facile modification of collagen directed by collagen mimetic peptides, J Am Chem Soc 127 (2005) 4130–4131.

[21] Y. Li, D. Ho, H. Meng, T.R. Chan, B. An, H. Yu, et al., Direct detection of collagenous proteins by fluorescently labeled collagen mimetic peptides, Bioconjug Chem 24 (2012) 9–16.

[22] Y. Li, C.A. Foss, D.D. Summerfield, J.J. Doyle, C.M. Torok, H.C. Dietz, et al., Targeting collagen strands by photo-triggered triple-helix hybridization, Proc Natl Acad Sci USA 109 (2012) 14767–14772.

[23] E.F. Banwell, E.S. Abelardo, D.J. Adams, M.A. Birchall, A. Corrigan, A.M. Donald, et al., Rational design and application of responsive α-helical peptide hydrogels, Nat Mater 8 (2009) 596–600.

[24] D. Papapostolou, E.H.C. Bromley, C. Bano, D.N. Woolfson, Electrostatic control of thickness and stiffness in a designed protein fiber, J Am Chem Soc 130 (2008) 5124–5130.

[25] J.M. Fletcher, R.L. Harniman, F.R.H. Barnes, A.L. Boyle, A. Collins, J. Mantell, et al., Self-assembling cages from coiled-coil peptide modules, Science 340 (2013) 595–599.

[26] H. Dong, S.E. Paramonov, L. Aulisa, E.L. Bakota, J.D. Hartgerink, Self-assembly of multi-domain peptides: Balancing molecular frustration controls conformation and nanostructure, J Am Chem Soc 129 (2007) 12468–12472.

[27] J.B. Matson, C.J. Newcomb, R. Bitton, S.I. Stupp, Nanostructure-templated control of drug release from peptide amphiphile nanofiber gels, Soft Matter 8 (2012) 3586–3595.

[28] J. Boekhoven, S.I. Stupp, Supramolecular materials for regenerative medicine, Adv Mat 26 (2014) 1642–1659.

[29] J.D. Hartgerink, E. Beniash, S.I. Stupp, Peptide-amphiphile nanofibers: A versatile scaffold for the preparation of self-assembling materials, Proc Natl Acad Sci USA 99 (2002) 5133–5138.

[30] D.T. Bong, T.D. Clark, J.R. Granja, M.R. Ghadiri, Self-assembling organic nanotubes, Angew Chem Int Ed 40 (2001) 988–1011.

[31] J.D. Hartgerink, J.R. Granja, R.A. Milligan, M.R. Ghadiri, Self-assembling peptide nanotubes, J Am Chem Soc 118 (1996) 43–50.

[32] R. Ketchem, W. Hu, T. Cross, High-resolution conformation of gramicidin A in a lipid bilayer by solid-state NMR, Science 261 (1993) 1457–1460.

[33] M.R. Ghadiri, J.R. Granja, L.K. Buehler, Artificial transmembrane ion channels from self-assembling peptide nanotubes, Nature 369 (1994) 301–304.

[34] M.R. Ghadiri, J.R. Granja, R.A. Milligan, D.E. McRee, N. Khazanovich, Self-assembling organic nanotubes based on a cyclic peptide architecture, Nature 366 (1993) 324–327.

[35] N. Khazanovich, J.R. Granja, D.E. McRee, R.A. Milligan, M.R. Ghadiri, Nanoscale tubular ensembles with specified internal diameters. Design of a self-assembled nanotube with a 13-Å. Pore, J Am Chem Soc 116 (1994) 6011–6012.

[36] J. Montenegro, C. Vazquez-Vazquez, A. Kalinin, K.E. Geckeler, J.R. Granja, Coupling of carbon and peptide nanotubes, J Am Chem Soc 136 (2014) 2484–2491.

[37] N.R. Zaccai, B. Chi, A.R. Thomson, A.L. Boyle, G.J. Bartlett, M. Bruning, et al., A de novo peptide hexamer with a mutable channel, Nat Chem Biol 7 (2011) 935–941.

[38] G. von Maltzahn, S. Vauthey, S. Santoso, S. Zhang, Positively charged surfactant-like peptides self-assemble into nanostructures, Langmuir 19 (2003) 4332–4337.

[39] R. Ischakov, L. Adler-Abramovich, L. Buzhansky, T. Shekhter, E. Gazit, Peptide-based hydrogel nanoparticles as effective drug delivery agents, Bioorg Med Chem 21 (2013) 3517–3522.

[40] M. Reches, E. Gazit, Controlled patterning of aligned self-assembled peptide nanotubes, Nat Nanotechnol 1 (2006) 195–200.

[41] A. Mishra, K.-H. Chan, M.R. Reithofer, C.A.E. Hauser, Influence of metal salts on the hydrogelation properties of ultrashort aliphatic peptides, RSC Adv 3 (2013) 9985–9993.

[42] Y. Luo, M.S. Shoichet, A photolabile hydrogel for guided three-dimensional cell growth and migration, Nat Mater 3 (2004) 249–253.

[43] B.D. Ratner, S.J. Bryant, Biomaterials: Where we have been and where we are going, Annu Rev Biomed Eng 6 (2004) 41–75.

[44] S.P. Massia, J.A. Hubbell, An RGD spacing of 440nm is sufficient for integrin IS $\alpha_v\beta_3$-mediated fibroblast spreading and 140nm for focal contact and stress fiber formation, J Cell Bio 114 (1991) 1089–1100.

[45] H. Storrie, M.O. Guler, S.N. Abu-Amara, T. Volberg, M. Rao, B. Geiger, S.I. Stupp, Supramolecular crafting of cell adhesion, Biomaterials 28 (2007) 4608–4618.

[46] M. Zhou, A.M. Smith, A.K. Das, N.W. Hodson, R.F. Collins, R.V. Ulijn, J.E. Gough, Self-assembled peptide-based hydrogels as scaffolds for anchorage-dependent cells, Biomaterials 30 (2009) 2523–2530.

[47] Z.L. Luo, Y.Y. Yue, Y.F. Zhang, X. Yuan, J.P. Gong, L.L. Wang, et al., Designer D-form self-assembling peptide nanofiber scaffolds for 3-dimensional cell cultures, Biomaterials 34 (2013) 4902–4913.

[48] M. Ni, W.H. Tong, D. Choudhury, N.A.A. Rahim, C. Iliescu, H. Yu, Cell culture on MEMS platforms: A review, Int J Mol Sci 10 (2009) 5411–5441.

[49] S.M. Martin, R. Ganapathy, T.K. Kim, D. Leach-Scampavia, C.M. Giachelli, B.D. Ratner, Characterization and analysis of osteopontin-immobilized poly(2-hydroxyethyl methacrylate) surfaces, J Biomed Mater Res A 67A (2003) 334–343.

[50] W.Y. Seow, C.A.E. Hauser, Tunable mechanical properties of ultrasmall peptide hydrogels by crosslinking and functionalization to achieve the 3D distribution of cells, Adv Healthc Mater 2 (2013) 1219–1223.

[51] J. Song, E. Saiz, C.R. Bertozzi, A new approach to mineralization of biocompatible hydrogel scaffolds: An efficient process toward 3-dimensional bonelike composites, J Am Chem Soc 125 (2003) 1236–1243.

[52] J.D. Hartgerink, E. Beniash, S.I. Stupp, Self-assembly and mineralization of peptide-amphiphile nanofibers, Science 294 (2001) 1684–1688.

[53] T.D. Sargeant, C. Aparicio, J.E. Goldberger, H.G. Cui, S.I. Stupp, Mineralization of peptide amphiphile nanofibers and its effect on the differentiation of human mesenchymal stem cells, Acta Biomater 8 (2012) 2456–2465.

[54] A. Mata, Y.B. Geng, K.J. Henrikson, C. Aparicio, S.R. Stock, R.L. Satcher, et al., Bone regeneration mediated by biomimetic mineralization of a nanofiber matrix, Biomaterials 31 (2010) 6004–6012.

[55] Z.-Y. Lin, Z.-X. Duan, X.-D. Guo, J.-F. Li, H.-W. Lu, Q.-X. Zheng, et al., Bone induction by biomimetic PLGA-(PEG-ASP)n copolymer loaded with a novel synthetic BMP-2-related peptide in vitro and in vivo, J Control Rel 144 (2010) 190–195.

[56] M.P. Lutolf, F.E. Weber, H.G. Schmoekel, J.C. Schense, T. Kohler, R. Muller, et al., Repair of bone defects using synthetic mimetics of collagenous extracellular matrices, Nat Biotechnol 21 (2003) 513–518.

[57] A. Horii, X. Wang, F. Gelain, S. Zhang, Biological designer self-assembling peptide nanofiber scaffolds significantly enhance osteoblast proliferation, differentiation and 3-D migration, PLoS ONE 2 (2007) e190.

[58] American Burn Association. Burn Incidence and Treatment in the United States: 2013 Fact Sheet, <http://www.ameriburn.org/resources_factsheet.php>; 2013.

[59] E.S. Place, N.D. Evans, M.M. Stevens, Complexity in biomaterials for tissue engineering, Nat Mater 8 (2009) 457–470.

[60] Y. Loo, Y.-C. Wong, E.Z. Cai, C.-H. Ang, A. Raju, A. Lakshmanan, et al., Ultrashort peptide nanofibrous hydrogels for the acceleration of healing of burn wounds, Biomaterials 35 (2014) 4805–4814.

[61] M.R. Reithofer, A. Lakshmanan, A.T.K. Ping, J.M. Chin, C.A.E. Hauser, In situ synthesis of size-controlled, stable silver nanoparticles within ultrashort peptide hydrogels and their antibacterial properties, Biomaterials 35 (2014) 7535–7542.

[62] R. Langer, J.P. Vacanti, Tissue Engeering, Science 260 (1993) 920–926.

[63] R.N. Shah, N.A. Shah, M.M.D.R. Lim, C. Hsieh, G. Nuber, S.I. Stupp, Regenerative medicine special feature: Supramolecular design of self-assembling nanofibers for cartilage regeneration, Proc Natl Acad Sci USA 107 (2010) 3293–3298.

[64] J. Kisiday, M. Jin, B. Kurz, H. Hung, C. Semino, S. Zhang, A.J. Grodzinsky, Self-assembling peptide hydrogel fosters chondrocyte extracellular matrix production and cell division: Implications for cartilage tissue repair, Proc Natl Acad Sci USA 99 (2002) 9996–10001.

[65] S.L. Murphy, J. Xu, K.D. Kochanek, Deaths: Preliminary Data for 2010, Natl Vital Stat Rep 60 (2012) 1–51.

[66] C.E. Semino, J.R. Merok, G.G. Crane, G. Panagiotakos, S.G. Zhang, Functional differentiation of hepatocyte-like spheroid structures from putative liver progenitor cells in three-dimensional peptide scaffolds, Differentiation 71 (2003) 262–270.

[67] G. Mehta, C.M. Williams, L. Alvarez, M. Lesniewski, R.D. Kamm, L.G. Griffith, Synergistic effects of tethered growth factors and adhesion ligands on DNA synthesis and function of primary hepatocytes cultured on soft synthetic hydrogels, Biomaterials 31 (2010) 4657–4671.

[68] Centers for Disease Control and Prevention. National diabetes fact sheet: national estimates and general information on diabetes and prediabetes in the United States, 2011, <http://www.cdc.gov/diabetes/pubs/pdf/ndfs_2011.pdf>; 2011.

[69] S. Khan, S. Sur, C.J. Newcomb, E.A. Appelt, S.I. Stupp, Self-assembling glucagon-like peptide 1-mimetic peptide amphiphiles for enhanced activity and proliferation of insulin-secreting cells, Acta Biomater 8 (2012) 1685–1692.

[70] D.-J. Lim, S.V. Antipenko, J.B. Vines, A. Andukuri, P.T.J. Hwang, N.T. Hadley, et al., Improved MIN6 β-cell function on self-assembled peptide amphiphile nanomatrix inscribed with extracellular matrix-derived cell adhesive ligands, Macromol Biosci 13 (2013) 1404–1412.

[71] G. Edlin, E. Golanty, Health & Wellness, 11th ed., Jones & Bartlett Learning, (2012).

[72] A. Andukuri, W.P. Minor, M. Kushwaha, J.M. Anderson, H.-W. Jun, Effect of endothelium mimicking self-assembled nanomatrices on cell adhesion and spreading of human endothelial cells and smooth muscle cells, Nanomed Nanotechnol Bio Med 6 (2010) 289–297.

[73] A. Andukuri, Y.-D. Sohn, C.P. Anakwenze, D.-J. Lim, B.C. Brott, Y.-S. Yoon, et al., Enhanced human endothelial progenitor cell adhesion and differentiation by a bioinspired multifunctional nanomatrix, Tissue Eng C 19 (2012) 375–385.

[74] M.T. McClendon, S.I. Stupp, Tubular hydrogels of circumferentially aligned nanofibers to encapsulate and orient vascular cells, Biomaterials 33 (2012) 5713–5722.

[75] L.W. Chow, L.-J. Wang, D.B. Kaufman, S.I. Stupp, Self-assembling nanostructures to deliver angiogenic factors to pancreatic islets, Biomaterials 31 (2010) 6154–6161.

[76] S. Ghanaati, M.J. Webber, R.E. Unger, C. Orth, J.F. Hulvat, S.E. Kiehna, et al., Dynamic in vivo biocompatibility of angiogenic peptide amphiphile nanofibers, Biomaterials 30 (2009) 6202–6212.

[77] S. Sur, E.T. Pashuck, M.O. Guler, M. Ito, S.I. Stupp, T. Launey, A hybrid nanofiber matrix to control the survival and maturation of brain neurons, Biomaterials 33 (2012) 545–555.

[78] T.C. Holmes, S. de Lacalle, X. Su, G.S. Liu, A. Rich, S.G. Zhang, Extensive neurite outgrowth and active synapse formation on self-assembling peptide scaffolds, Proc Natl Acad Sci USA 97 (2000) 6728–6733.

[79] S. Koutsopoulos, S.G. Zhang, Long-term three-dimensional neural tissue cultures in functionalized self-assembling peptide hydrogels, Matrigel and Collagen I, Acta Biomater 9 (2013) 5162–5169.

[80] C.E. Schmidt, J.B. Leach, Neural tissue engineering: Strategies for repair and regeneration, Annu Rev Biomed Eng 5 (2003) 293–347.

[81] P. Ducheyne, R.L. Mauck, D.H. Smith, Biomaterials in the repair of sports injuries, Nat Mater 11 (2012) 652–654.

[82] G.A. Silva, C. Czeisler, K.L. Niece, E. Beniash, D.A. Harrington, J.A. Kessler, et al., Selective differentiation of neural progenitor cells by high-epitope density nanofibers, Science 303 (2004) 1352–1355.

[83] M. Iwasaki, J.T. Wilcox, Y. Nishimura, K. Zweckberger, H. Suzuki, J. Wang, et al., Synergistic effects of self-assembling peptide and neural stem/progenitor cells to promote tissue repair and forelimb functional recovery in cervical spinal cord injury, Biomaterials 35 (2014) 2617–2629.

[84] S. Sur, C.J. Newcomb, M.J. Webber, S.I. Stupp, Tuning supramolecular mechanics to guide neuron development, Biomaterials 34 (2013) 4749–4757.

[85] A.M. Hung, S.I. Stupp, Simultaneous self-assembly, orientation, and patterning of peptide-amphiphile nanofibers by soft lithography, Nano Lett 7 (2007) 1165–1171.

[86] A. Mata, L. Hsu, R. Capito, C. Aparicio, K. Henrikson, S.I. Stupp, Micropatterning of bioactive self-assembling gels, Soft Matter 5 (2009) 1228–1236.

[87] L. Adler-Abramovich, E. Gazit, Controlled patterning of peptide nanotubes and nanospheres using inkjet printing technology, J Pept Sci 14 (2008) 217–223.

[88] R. Langer, N.A. Peppas, Advances in biomaterials, drug delivery, and bionanotechnology, AIChE J 49 (2003) 2990–3006.

[89] S. Koutsopoulos, L.D. Unsworth, Y. Nagai, S. Zhang, Controlled release of functional proteins through designer self-assembling peptide nanofiber hydrogel scaffold, Proc Natl Acad Sci USA 106 (2009) 4623–4628.

[90] M.J. Webber, J.B. Matson, V.K. Tamboli, S.I. Stupp, Controlled release of dexamethasone from peptide nanofiber gels to modulate inflammatory response, Biomaterials 33 (2012) 6823–6832.

[91] M.R. Reithofer, K.-H. Chan, A. Lakshmanan, D.H. Lam, A. Mishra, B. Gopalan, et al., Ligation of anti-cancer drugs to self-assembling ultrashort peptides by click chemistry for localized therapy, Chem Sci 5 (2014) 625–630.

[92] A. Lakshmanan, D.W. Cheong, A. Accardo, E. Di Fabrizio, C. Riekel, C.A.E. Hauser, Aliphatic peptides show similar self-assembly to amyloid core sequences, challenging the importance of aromatic interactions in amyloidosis, Proc Natl Acad Sci USA 110 (2013) 519–524.

Fabrication of Drug Delivery Systems Using Self-Assembled Peptide Nanostructures

Daniel Keith, Honggang Cui
Department of Chemical and Biomolecular Engineering and Institute for NanoBioTechnology, Johns Hopkins University, Baltimore, Maryland

5.1 INTRODUCTION

The development of drug delivery systems has become an area of intense research focus in the past few decades. The promise of a future where nanomedicine allows for personalized treatment, as well as much-improved therapeutic and diagnostic efficacy at the cellular and molecular level, is powerful motivation for the conquest of the nano–bio interface. A large variety of methods and approaches have emerged in the fabrication of nanoscale therapeutics; unfortunately, only a select few have successfully made their way through clinical trials to become approved medical therapies. Be that as it may, the field of drug delivery remains highly optimistic, with a great potential to revolutionize medicine. In this chapter, we will analyze the criteria for an effective drug-delivery system, and the potential of self-assembling peptide-based nanostructures to meet these challenges. This chapter also outlines the details for the rational design of these drug systems, as well as current examples of cutting-edge work that harness such designs.

5.2 DRUG DELIVERY SYSTEM DESIGN

5.2.1 Tumor Targeting

Chemotherapy is the only feasible way to cure metastatic cancer that has already spread from the primary site. Historically, anticancer drugs are small molecules that are effective at killing cancer cells and are the focal point of cancer treatments (in addition to surgery and radiation). However, most anticancer drugs are very cytotoxic, and for cell-cycle-specific-compounds, their mechanism of action is to kill cancerous cells with a slightly higher preference compared to other proliferating healthy

Micro and Nanofabrication Using Self-Assembled Biological Nanostructures. DOI: 10.1016/B978-0-323-29642-7.00005-9

cells [1]; this often results in devastating side effects for the patients, and can even prove fatal. This necessitates the development of strategies for targeted therapies.

Nanotechnology formulations offer increased pharmacokinetic capability by means of both passive and active targeting. Passive targeting takes advantage of the size and surface chemistry of the nanoparticles (NPs), which permits them to circulate a longer period of time and eventually extravasate through the leaky endothelial wall adjacent to tumor tissues through the enhanced permeability and retention (EPR) effect [2–6]. Active targeting consists of NPs that contain molecules (e.g., peptides or ligands) that bind to specific biomarkers at target tissues [1]. These target ligands usually bind to cellular membrane receptors that are known to have high expression levels in cancer tissues [7]. Thus, NPs can theoretically be engineered to preferentially accumulate in tumor vasculature.

In spite of this, the development of nanotherapeutics has struggled to overcome the multiple physiological barriers and the widely diverse nature of tumor tissues that could vary from patient to patient [8]; the mechanism of tumor heterogeneity is still, unfortunately, poorly understood [9]. Many challenges associated with this heterogeneity often involve different biological barriers that end in the same result: the drug delivery system does not provide a sufficient bioavailable and homogeneous dose to its target [8].

5.2.2 Desirable Carrier Characteristics

NPs intended for therapeutic use, in order to effectively enhance their carriers, must be inherently multifaceted in their design [10]. In addition to transporting their drug cargo to their respective targets, they must also be able to enter the cells and dissociate to ensure their cargo's viability and bioavailability. Thus, efforts to engineer functional therapeutic nanocarriers must incorporate several functions to successfully achieve the desired clinical result.

In order to effectively overcome such obstacles through systemic or local treatments, it is useful to define the needed qualities that nanotherapeutics must possess. Ideally, drug nanocarriers will have improved half-life in the bloodstream when systemically delivered, and will not exhibit any "burst" release (an unfortunately common phenomenon wherein drugs dissociate prematurely from their carriers and fail to reach the target tissue in a significant dose). Furthermore, the carrier should also possess the versatility to be able to effectively release its drug at the target site, ensuring high bioavailability and intracellular uptake.

It is paramount that the drug carrier has a mechanism to avoid immune detection and consequent opsonization and clearance from the body. In fact, one of the biggest obstacles in the approval of a nanotherapeutic is demonstration of its low immunogenicity [11]. Compared with their non-modified drug counterparts, many nano-encapsulated drugs have yielded lower instances of hypersensitivity in patients. In the ideal case, nanocarriers act as a shield to lower their drug's antigenicity to prevent its opsonization by the immune system. In contrast to this capability, it is critical for the carrier to then "stick to," and be rapidly internalized by, cancer cells upon penetrating the tumor vasculature [12]. The function of the carrier in this case is to mediate or expedite this process in comparison to administration of free cargo alone, often by means of enhanced permeability and retention.

However, although tumor localization through the EPR effect is a great advantage of targeting NPs, the primary advantage in using drug-delivering nanocarriers is their increased accumulation and uptake in cancer cells [13,14]. For this reason, when designing a therapeutic, it is important to determine the specific type of cell or tissue that is being targeted; this proves quite the challenge when considering the unique microenvironments of solid tumors, which have a wide distribution of cell types [11]. This points to the need to program targeting nanotherapeutics to work on multiple fronts, so that all of the appropriate physiological and pharmacokinetic barriers can be simultaneously hurdled to ensure proper therapeutic efficacy. Therefore, it is evident that the design of novel drug-delivery systems must strive to optimize performance on an all-encompassing level in order to achieve maximum potency (Figure 5.1).

5.2.3 Drug Loading Strategies

Drug delivery systems, though aimed to improve the same set of pharmacokinetic traits for given therapeutic materials, have been extensively researched in a myriad of forms, their compositions varying as much

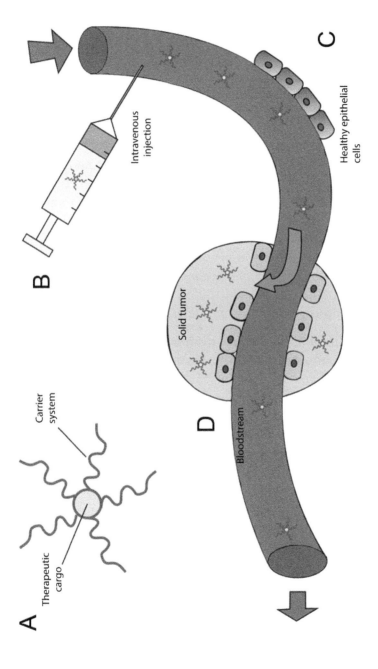

Fig. 5.1. Schematic illustration of the basic principles of systemic drug-delivering nanoparticles. (A) Cargo is either encapsulated by, or conjugated to, its carrier; (B) it is then administered intravenously. (C) The tight vasculature of healthy epithelial cells prevents the nanoparticles from entering healthy tissues, allowing for circulation in the bloodstream. (D) When the drug encounters "leaky" tumor vasculature, it is able to traverse the gaps in the cancerous epithelium via the EPR effect, thus accumulating in solid tumors.

as their mechanisms of action. This is to be expected, since therapeutic agents and their purpose are themselves quite diverse. A wide range of nanostructures, including carbon nanotubes (NTs), buckyballs, liposomes, inorganic NPs (such as heavy metals), dendrimers, quantum dots, micelles, and other natural and synthetic molecules have been investigated as platforms for drug delivery systems.

In general, drugs assume two modes of association with their carriers: physical encapsulation or chemical conjugation. The former method consists of synthesizing molecular structures that can physically internalize a therapeutic agent, stabilizing it via storage inside an internal cavity and/or through noncovalent interactions between drug and carrier [15]. This can be readily accomplished with micelles, or porous NPs, wherein the encapsulating structure has a net hydrophobicity to stabilize the entrapped molecule [16]. Shen et al. discussed a select number of methods for the optimization of these physically encapsulating effects when engineering micelles, such as polymeric coating for immune-stealth and multiple carrier shells for lower diffusion rates [12]. The second method of forming drug delivery systems is by means of direct chemical conjugation of the cargo molecule to the carrier. This strategy exhibits several advantages in that it confers greater control over the triggered release of the therapeutic through a biochemically cleavable linker [12,15,17]. Just as a physically encapsulating carrier must be rationally designed such that it can both successfully load its cargo and release it in the appropriate spatiotemporal manner, chemically conjugated therapeutics must also have a means of dissociation from their nanocarriers. Failure of the carrier to be cleaved from its cargo inhibits the drug's bioactivity; if this effect is too prevalent, it may even reduce the efficacy of the therapeutic to a lower level than that of the free drug [18]. Alternatively, a drug that is cleaved too quickly from its carrier will exhibit the aforementioned burst release and will be rapidly cleared from the body [19].

5.2.4 Transition to Practicality: Safety and Scale-Up

Nanotherapeutics have also been slow to make their way into approved medical treatment regimens, because their complexity increases the necessity of their thorough evaluation on multiple fronts [1]. Their physiochemical properties, biodistribution within the body, and toxicity levels must be quantitatively analyzed via both *in vitro* and *in vivo* testing. As such, their complex nature has caused their preclinical evaluation to become a rate-limiting step in their regulation and scale-up [1]. One major component in the proper analysis of NPs is a review of their new pharmacokinetic properties, which are not guaranteed to be comparable to their unmodified ancestors [11].

Administration of a drug coupled to a nanocarrier, by necessity of the carrier's purpose, changes the medicine's mechanism of systemic retention, targeting, and uptake. NPs also have a very high surface area-to-volume ratio by nature of their size, which gives them quite unusual chemical and physical characteristics. These substances have the potential to be toxic in ways not fully understood [20]. In summation, the field of nanomedicine holds great promise, but is currently hindered from advancing through clinical trials and expanding its influence because of the large learning curve currently associated with the characterization of its safety.

The scale-up transition from test-tube quantities to industrial-scale batches of chemicals is also a daunting concept. Current research may demonstrate proof-of-concept with milligram quantities, yet this remains massively far away from the successful, safe, and consistent production of high-quality material with any sort of profitable yield [11]. Indeed, it will be a major focus of bioengineers to efficiently manage the translation of this technology into more conventional biomedical utilization.

The vast range of drug delivery vesicles studied for biomedical application that rely on both of these encapsulation types are thoroughly documented in a massive host of scientific literature; to explore all of them thoroughly is both beyond the scope and tangent to the point of this written work. One of the most promising areas of research in drug delivery and tissue engineering is the utilization of peptides – either physically via encapsulation into self-assembled nanostructures or through chemical conjugation to bestow self-assembling capabilities – as a biodegradable, physiologically sensitive, inherently "tunable," and remarkably facile design platform for highly sophisticated drug delivery systems. In the next sections, this chapter will investigate the self-assembly of these peptides, the unique utility and power of this phenomenon in the construction of relevant biological nanostructures [21], how such behavior forms discrete, meta-stable nanoarchitectures, and reports of cutting-edge applications of these drug amphiphiles at the nano–bio interface.

5.3 PEPTIDE-BASED NANOSTRUCTURES AS DRUG CARRIERS

Peptides offer several molecular attributes that enable their unique use for drug delivery. First of all, the most intriguing properties that favor the use of peptide-based materials as drug carriers are their inherent biocompatibility and biodegradability. Importantly, the degradation of peptides is mediated by proteases, and is further tunable by the incorporation of specific amino acid sequences or control over their self-assembled structures. Upon finishing their duty as carriers, peptides will serve as nutrients to cells. Second, peptides are excellent molecular building blocks for creating well-defined nanostructures of any size and shape. In particular, β-sheet-forming peptides demonstrate the extraordinary ability to assemble into 1D nanostructures through intermolecular hydrogen bonding. The chemical design versatility of peptides, in combination with their ability to adopt specific secondary structures, provides a unique platform for the design of materials with controllable structural features at the nanoscale. Third, peptides can be designed to be bioactive and to mediate many important biological events. The use of bioactive epitopes to improve the bioactivity of NPs is a widely adopted strategy. Last, oligopeptides can be produced rather easily in large quantities using standard solid-phase synthesis protocols at a relatively low cost. Current developments in the bacterial synthesis of peptides can further lower the cost, making it suitable for large-scale industrial production.

5.3.1 Primary and Secondary Structure – Their Relation to Self-Assembly Basics

At its smallest length scale, peptide-based amphiphile design is dictated by the molecule's primary structure. The dynamics of the assembled nanostructure, including its size, shape, and bioactivity of its surface chemistry, are directly dependent upon the primary sequence of amino acids in its peptide segment. The protonation and deprotonation of the respective carboxylate or ammonium groups on amino acids have an inherent "tunability" [22], a trait that allows for constructing pH-responsive nanostructures and materials. Because the emergent functionality of self-assembling entities is determined by the design of its component building blocks [21], the informed design of a peptide's primary sequence to obtain thermodynamically favorable patterns in secondary structure will yield meaningful results in the consequently formed nanoarchitectures.

Peptides have been recognized as a basis for building self-assembling nanostructures that can organize in up to three dimensions to produce hierarchically ordered networks [22], thanks to the noncovalent interactions within their secondary structure. Specifically, peptides whose amino acid side-chains are likely to form β-sheet secondary structures have demonstrated excellent capability to form 1D self-assembling discrete nanoarchitectures [23]. Interesting examples include peptide fibrils, cylindrical nanofibers, helical fibers, twisted ribbons, helical ribbons, nanobelts, and NTs. Access to this broad range of 1D morphologies is a direct consequence of the design versatility of the peptide primary structure. The spontaneous formation of such structures are driven by both van der Waals forces in the hydrophobic segments as well as hydrogen bonding and electrostatic interactions between the adjacent amino acid side chains on the periphery of the formed shapes; a balance must exist among the electrostatic repulsive forces between the charged surface amino acids and the hydrophobic effect on the inner sequence [22,23]. With careful consideration of such properties, the rational design of these structures holds immense promise in the controlled fabrication of hydrogels and higher-order structures for biomedical utilization. Therefore, self-assembly of rationally designed peptidic building units is a remarkable tool in the production of customizable engineered biomaterials at the nanoscale. To echo previous points, this warrants further exploration for therapeutic and diagnostic use.

A very important class of peptide-based materials are the high-aspect-ratio peptide amphiphile (PA) nanofibers designed by Stupp Group that were found to have remarkable bioactivities [22]. A representative PA design is illustrated in Figure 5.2, possessing four key structural features. Region 1, the hydrophobic domain, consists of a long alkyl tail or a hydrophobic anticancer drug, or any other hydrophobic units depending on the purpose of the application. Region 2 comprises hydrophobic

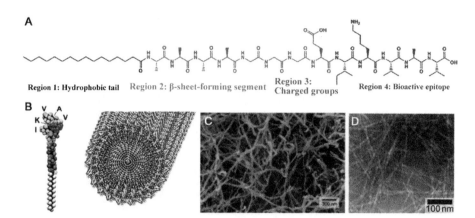

Fig. 5.2. (A) Molecular structure of a representative PA with four rationally designed chemical entities. (B) Molecular graphics illustration of an IKVAV-containing PA molecule and its self-assembly into nanofibers; (C) scanning electron micrograph of the IKVAV nanofiber network formed by adding cell media to the PA aqueous solution; (D) transmission electron micrograph of the IKVAV nanofibers. From Ref. [22].

amino acids that have a strong propensity to form intermolecular hydrogen bonding. This unique feature leads to the 1D nature of the resulting self-assembled nanostructures. Region 3 contains charged amino acids for enhanced solubility in water and for the design of pH- and salt-responsive nanostructures and networks. The number of charged amino acids added should be enough to ensure good solubility and assist with peptide purification, but not too many so as to interfere with the self-assembly of PAs into 1D nanostructures under physiological conditions. Region 4 is typically used for the presentation of bioactive signals, usually through epitopes that can interact with cells or proteins for cell signaling or tumor targeting. The internal packing of these PA molecules into high-aspect-ratio nanofibers is illustrated in Figure 5.2B.

Tirrell et al. have designed and developed a series of water-soluble peptide amphiphile micelles [24,25]. These protein-analogous micelles provide a multivalent display of functional peptides that may find use in a wide range of therapeutic and diagnostic settings. Tirrell lab have examined PAs of variable alkyl tail lengths that contained a collagen-based peptide head group; these amphiphiles, designed to be trypsin-cleavable, showed higher degradation rates when forming spherical micelles compared to when they aggregated into membrane-like bilayers [26]. In addition, the type of structure formed could be determined by altering the alkyl tail length. Elongation of the hydrocarbon segment causes a micelle-to-bilayer aggregate shift as a consequence of the way their volume is distributed; because spherical geometries contain most of their volume at the surface, alkyl chains must bend more to satisfy this spatial requirement. Longer carbon chains are more preferential to low-curving states in layered aggregates. In fact, in 14- and 16-carbon-containing PAs, higher-order disc-shaped micelles were observed to stack to form fibrous higher-order structures [26].

5.3.2 Biodegradability of Peptide Carriers

Just as in the development of any other therapeutics, it is necessary to consider the safety of the compounds being used. The pharmacokinetic profile of a drug delivery system should be investigated, but perhaps just as important as the cargo's ultimate fate is that of the carrier; the degradation products of the delivery vector should be known [12]. Covalent conjugation of peptides to therapeutic cargo causes little concerns here, since after the target site is reached, it is inherently biodegradable by cellular proteases into simple amino acids [27]. As the building blocks of proteins, by-product amino acids are of no more danger to a cell than the enzymes that give it life. Such obvious compatibility with biological systems has lent to its ready incorporation into therapeutic nanostructures [17,27].

One important consideration in the design of peptide-based drug delivery system is the selective degradation of the resulting supramolecular materials by enzymes/proteases of interest, because chemical breakdown presents the first step toward ultimate clearance of these synthetic peptides once their duties are finished, and also because controlled degradation by relevant enzymes regulates the release rate of the delivered cargo. There has been a rapidly growing interest in the development of peptide-based supramolecular filaments responsive to specific matrix metalloproteases (MMPs) due to their important biomedical applications. Examples include the RADA peptides containing MMP-2-specific substrates by Chau et al. [28], the self-assembling multidomain peptides developed by the Hartgerink lab (responsive to MMP-2) [29], and the self-assembling β-hairpin degrading peptides developed by Schneider, Pochan, and Giano (responsive to MMP-13) [30]. Cui et al. reported a cross-linking strategy to construct supramolecular filaments that can specifically break down in the presence of MMP-2 [31]. Through linkage of an oligoproline segment to an amyloid-derived peptide sequence, an amphiphilic peptide was designed to undergo a rapid morphological transition in response to pH variations. MMP-2-specific peptide substrates were used as multivalent cross-linkers to covalently fix the amyloid-like filaments in the self-assembled state at pH 4.5. They showed that the cleavage reaction actually takes place more efficiently on the degradable linkers that joined the assembled nanostructures, leading to the breakdown of cross-linked filaments.

5.3.3 Usage of Bioactive Peptides in Carriers

Considering the facile incorporation of anticancer drugs into peptide-based nanostructures, the nature of the peptide's bioactivity becomes important. Previously, it was discussed that the peptide's primary structure should be the focus of the design, for it directly determines the nature of the nanoarchitectures. This remains true, but now for another reason: the amino acid sequence in the nanocarrier can be tailored to emulate the functions of biomolecules such as proteins [16,22,32]. Peptides that act as target ligands or epitopes for specific cell receptors or extracellular matrix (ECM) elements can be linked to a small-molecule drug to improve its uptake in cells preferential to binding that sequence. One of the most frequently analyzed and used epitopes as a target ligand is RGDS, a common amino acid sequence that binds to extracellular fibronectins and readily facilitates ligand–cell adhesion [32,33]. In addition to this, others have synthesized peptide nanocarriers that use ligands more specific for certain tumor types. Tumors will contain specific biomarkers in their microenvironments that readily interact with ligands such as transferrin, epidermal and other growth factors, folate, and certain glycoproteins [34,35]. For instance, Tirrell et al. reported the usage of CREKA, a clot-binding peptide, in fluorescently labeled protein-analogous micelles that can effectively deliver cargo across the blood–brain barrier to treat glioblastoma [16]. This development holds great promise for peptides as optimizable drug delivery vesicles, considering past difficulties in the delivery of therapeutics across the blood–brain barrier [36].

Related to the common RGD binding sequence is a recently discovered tumor-penetrating peptide, iRGD, which was reported to penetrate further into extravascular tumor tissue than most anticancer drugs [37]. Similar to its RGD counterpart, it first binds to overexpressed integrins on cancerous endothelial cells and is then proteolytically cleaved; the resultant shortened peptide, GRGDK/R, gains an affinity for Neuro-pilin-1, the binding of which triggers tissue penetration [37]. Ruoslahti et al. demonstrated the tumor-specific penetration activity of the iRGD peptide through coadminstration with several poorly penetrating substances, showing increased permeability of tumors and, in the case of small molecule drugs, increased therapeutic index. This is thought to happen through an active mechanism as opposed to passive uptake through the EPR effect, in light of the receptor and energy-dependence of the iRGD uptake kinetic profile [37]. This active uptake also occurs in a matter of minutes, whereas systemic uptake through EPR takes several hours. For the coadminstration of the tumor-homing peptide with a therapeutic, no chemical conjugation of the two is deemed necessary: the systemically delivered iRGD peptide will improve the permeability of the tumor to small molecule drugs without making a new chemical substance, allowing drugs to be administered on a bulk scale [37].

Just as recognizing fibrin depositions as characteristic of brain tumors suggested the usage of the CREKA pentapeptide in the aforementioned design, overexpressed receptors in cancer cells are a prime target for known bioactive peptide ligands. Torchilin recently discussed examples of several peptides as target ligands to deliver liposome-containing chemotherapy drugs to specific sites [38]. A specific example is the usage of the 28-amino acid mammalian neuropeptide, Vasoactive intestinal peptide (VIP), to actively target human breast cancer [39]; drug delivery vessels containing VIP show preferential internalization in cancerous tissues due to a fivefold increase in their VIP receptor (VIP-R) expression. Cell-penetrating peptides, another important class of bioactive peptides, are small and cationic (typically with a high number of positively charged lysines and arginines [40]), and have been incorporated into a variety of drug vehicles due to their known capacity to effectively transport cargo across cell membranes [41,42].

The utilization of bioactive peptides in drug delivery systems can improve the pharmacokinetic profile of the cargo not only by enhancing its passive or active targeting abilities, but also by aiding its systemic retention and resistance to immune detection. One example of this approach is the reported usage of peptides that intrinsically block phagocytosis by immune cells. Just as human antibodies recognize and mark foreign molecules via their antigenic binding sites, cells native to an organism contain "self" signals that identify it as an entity not to be targeted for immune response. Discher et al. recently discovered a "self" peptide similar to human CD47, which is recognized by immune receptor CD172a [43]. These "self" peptides have been shown to slow the rate of macrophage-mediated clearance of encapsulated nanotherapeutics as small as viruses, as well as enhance tumor targeting through enhanced permeability and retention. This provides a promising and novel platform for developing immune-stealthy drug delivery vesicles, as it is well reported that any polymer-based system can ultimately prove to be immunogenic [44]. When compared to PEG-coated nanobeads, which can prevent development of NP antigenicity [45,46] but ultimately result in opsonization after repeated exposure [43], hCD47-conjugated virus-sized particles displayed enhanced circulation time as a result of its inhibition of macrophage phagocytosis. With respect to their increased therapeutic index due to increased accumulation at tumor sites, a 16-to-22-fold enhancement was demonstrated in an *in situ* imaging study of hCD47-conjugated nanobeads [44]. Discher's design of drug delivery systems composed of "self" peptides is a powerful tool due to its stealth, retention, and targeting abilities; this monofunctionality is a desirable trait for all current and future therapies that the field of nanomedicine may produce.

Conversely, it may also be desirable in some cases to invoke an immune response, such as in the development of vaccines, where the presence of antigens promotes the proliferation of antibodies without causing any harm to the recipient. Tirrell et al. have also developed a peptide-based system with such immune activity. Their work investigates the usage of PA micelles as self-adjuvanting agents that can effectively present antigens to dendritic cells in order to induce cytokine production and therefore create an effective response from the immune system [47]. Theoretically, an effective peptide-based delivery system is able to concentrate and protect its presented antigen from degradation and increase its internalization by dendritic cells; this was demonstrated via the conjugation of two palmitic alkyl tails to a known T-cell epitope from ovalbumin, a tumor antigen [47]. This relatively facile design results in demonstrated self-assembly of NPs that have immunostimulatory capabilities, which is certainly deserving of attention from immunologists, bioengineers, and biomedical researchers alike.

In light of these examples, it is worth briefly discussing the emerging pattern in the development of bioactive peptide-based therapies. The fundamental starting point of this strategy, that certain peptide sequences are directly involved in biological transactions, must ultimately be informed by basic science. As more potent target ligand-based therapies develop, it will behoove biologists to further investigate interactions between ligands and receptors so as to present a new potential therapeutic target. As follows, the role of applied science and engineering will be to systematically develop appropriate therapeutic agents and treatment strategies that adequately harness this information. Efforts must be taken to ensure that the bioactivity of a peptide chosen for a specific biological purpose is not hampered by its incorporation into a drug delivery system.

5.3.4 Encapsulation of Drugs into Peptide Assemblies

In the construction of self-assembling peptides that would form supramolecular nanoarchitectures, the amphiphilic nature of the compound synthesized acts as a major driving force. These amphiphiles rely not only on the incorporation of hydrophilic peptides, but also upon hydrophobic building blocks, such as alkyl tails [22]. A modification to this strategy that is of great value is to consider the chemical conjugation of peptides as a natural candidate for application to anticancer drugs [48], which usually incur several problems due to their net hydrophobicity. Most chemotherapeutic agents are hydrophobic, small-molecule drugs that suffer from high toxicity and poor water solubility [1,48]; conjugation of these drugs to hydrophilic peptides would both create an amphiphilic system necessary for self-assembly and also potentially improve their pharmacokinetics via their incorporation into a drug-delivering nanocarrier. Indeed, such peptide-based drug delivery systems are currently of wide scientific interest, and have been reported to both increase the therapeutic index and reduce toxic side effects of many anticancer drugs [35,49,50]. Strategies in the rational design and implementation of such systems are elaborated upon in this section.

Nanostructures with high aspect ratios have demonstrated many advantages as drug carriers, such as longer half-life in systemic circulation and distribution [51,52] and increased target selectivity and uptake [53]. While micelles contain these traits, their design presents their own set of challenges in their consistent and tailorable synthesis and assembly

[54]. A more preferable design in this case is the usage of molecular building blocks that possess strong intermolecular attractions [55]. PAs with strong hydrogen-bonding capacity [22,33] and hydrophobic interactions within the cargo [24,56] are examples of such attractions. This dual capability suggests the potential to encapsulate hydrophobic anticancer drugs [55]. Unfortunately, previous usage of PAs as drug carriers has been subject to poor drug-loading capacities (DLC), around 2–5% [57] due to the high internal packing order of their hydrophobic carbon segments, limiting space for drug uptake [58]. To address these issues, Cui et al. incorporated multiple short hydrophobic tails to make the nanostructure's inner domain more enlarged for a potentially higher loading efficiency [55]. Proof-of-concept for the facile synthesis of poly-hydrocarbon-Tat nanofibers as carriers of the chemotherapeutic Paclitaxel (PTX) was shown to yield a drug loading of 7.0% by weight, a considerable gain from what has thus far been accomplished [55].

Through rigorous examination via cytotoxicity studies, the Tat nanofiber carriers showed efficient transport of their hydrophobic molecular cargo *in vitro* for various cancer cell lines by an active mechanism of internalization (not freely diffusing across the cell membrane). The encapsulated PTX showed similar IC_{50} values to free PTX in three studied cancer cell lines, demonstrating that the Tat nanofiber did not reduce efficacy; as in previously discussed studies, the nanocarrier itself was also demonstrated to not cause significantly high cytotoxicity by itself, confirming the conserved bioactivity of the delivered drugs [55]. Flow cytometry evaluation also confirmed a similar mechanism of action between the free drug and the prodrug [55]. These findings suggest that the Tat nanovector shows further potential for optimization, including the addition of a lysosomal escape ability, which would improve overall efficacy of the therapeutic. Finally, these findings suggest a shift of focus for the field of drug delivery, adding promise to drugs that have high potency inside cells but are deficient in their degree of cellular uptake and retention [55].

In sharp contrast to a great diversity of 1D nanostructures formed by self-assembly of peptide building units, it is rare and often challenging to construct vesicular morphologies using peptides as building block. Deming's lab successfully prepared polypeptide vesicles by conformation-specific assembly [59]. These synthetic vesicles can be tuned in size and shape by the ordered conformations of the polymer segments, in a manner that resembles capsid assembly in viruses [59]. Furthermore, these molecules can gain biofunctionality that allows them to be responsive to external stimuli, such as pH or temperature. This potentially enables them to act as biosensitive "switches" for drug delivery systems. The amphiphilicity of these peptide domains allows for more efficient control of these vesicle systems; the combination of structure and function that these peptides impart major advantages in the synthesis of encapsulated therapeutics [59].

Deming et al. further designed and developed polyarginine and polyleucine segment-based vesicles that are stable and can incorporate a large payload of both hydrophobic and water-soluble cargo alike (Figure 5.3) [60]. Remarkably, the polyarginine strands are

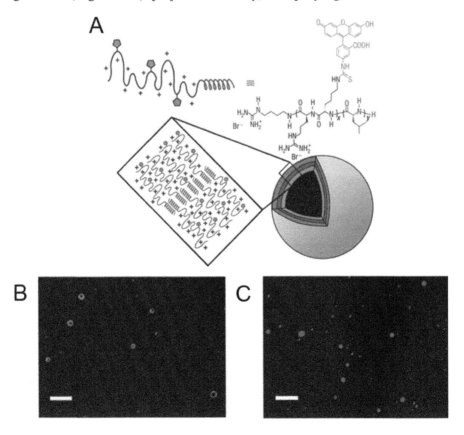

Fig. 5.3. (A) Schematic of a polyarginine ($R_{60}L_{20}$) vesicle from Deming's lab. (B) Confocal microscopy images of 1 μm extruded vesicles (scale bar 5 μm). (C) Confocal image of vesicles containing Texas-Red-labeled dextran. Adapted from Ref. [60].

multifunctional – they contribute to both the self-assembly of the vesicular structure as well as the mechanism of intracellular delivery of cargo. Whereas the large amount of positive charge in the polycations warrants concerns of high toxicity (as monomeric cations can cause cell lysis), Deming et al. demonstrated that similar to cation-based hydrogels, their self-assembly into such structures diminishes this hazard.

5.4 SELF-ASSEMBLING DRUG AMPHIPHILES AS NOVEL DELIVERY SYSTEMS

5.4.1 Incorporation of Anticancer Drugs as Molecular Building Blocks

In the self-assembling molecular nanocarrier strategy for the delivery of anticancer drugs, the drug component has typically been thought of only as cargo that must be delivered, and not as an additional building block to enhance the facility of self-assembly [61]. Recently, Cui et al. have reported a quantitative, high-capacity loading of anticancer drugs into potentially self-deliverable nanodiscs; folic acid, an important vitamin that is readily uptaken by cells and has recently been investigated as a target ligand for certain cancers [62], can be tuned with various solvent ratios to self-assemble into both nanofibers and microlozenges [61,63–65]. With more problematic drugs that do not readily assemble in aqueous solution, net hydrophilic peptides with β-sheet-forming side chains can be conjugated (Figure 5.4); the hydrophobic interaction of the cargo and the hydrogen bonding of the peptide chains contribute to the formation of stable 1D structures, such as nanofibers and NTs. This was the applied method for the anticancer drug Camptothecin (CPT), which was covalently linked to the β-sheet-forming Tau peptide. The Tau-conjugated CPT molecules can also be "tuned" to transition between different nanostructures [61], including NTs and nanofibers. The examination of these two compounds demonstrates the complex nature of small molecule drug assembly in general.

Most polymeric and other types of carriers show both difficulty in gaining both a high drug loading capacity as well as significantly inconsistent amount of drug loading or conjugation; in contrast, small molecule prodrugs are more "monodisperse," but are much more susceptible to quick clearance and inactivation [66]. Self-assembly into micelles is

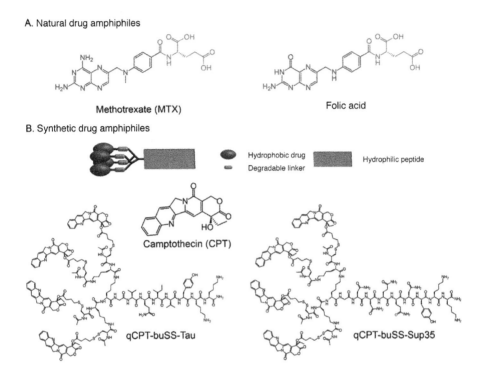

Fig. 5.4. Chemical structures of natural and synthetic drug amphiphiles used in this study. (A) Both methotrexate and folic acid contain a glutamic acid residue [marked in light gray (blue in the web version)] and can be regarded as amphiphilic molecules when deprotonated at a higher pH. (B) The creation of camptothecin (CPT) drug amphiphiles by the conjugation of four CPT molecules to one β-sheet forming peptide via a biodegradable linker. Two β-sheet forming sequences (VQIVYK and NNQQNY) were used to create two drug amphiphiles: qCPT-buSS-Tau and qCPT-buSS-Sup35. The linker used to bridge the drug and the peptide is responsive to glutathione, a reducing agent within cells. From Ref. [61].

possible [67–69], but is inherently limited in their structural tunability and drug-loading capacity. With these challenges in mind, Cui et al. have designed peptide-conjugated anticancer drug systems, consisting of CPT molecules bound in different proportion to β-sheet-forming "Tau" peptides through a degradable linker. These "drug amphiphiles" can readily self-assemble into stable supramolecular filaments [70] with flexible control of drug-loading capacity via alteration of the number of drug molecules conjugated per carrier. Circular dichroism measurements suggest the importance of the interaction of the hydrophobic drug molecules to initiate self-assembly of the amphiphiles into NTs [70], which can protect against premature cleavage of the bioactive cargo from its carrier and thus provide a mechanism of controlled, sustained release of drug [71]. Via a control, it was also shown that the Tau peptides used as carriers were not cytotoxic and provided no experimental artifacts; thus it can be concluded that the potency of the drug amphiphile is due to release of the drug cargo, and that the drug carrier will not indiscriminately kill cells [70]. The IC_{50} value of the di-CPT-Tau drug amphiphile was comparable to free CPT in a wide range of cancer cell lines during *in vitro* cytotoxicity studies. Thus, this discrete nanostructure with enhanced self-assembling capability from its cargo and a high and consistent drug loading provides insight into a new strategy for anticancer drug delivery.

Further specificity in the design of drug amphiphiles is gained through coupling of the cargo to different ends of the peptide carrier, a potentially very important design facet that has been widely unexplored [72]. Cui et al. have demonstrated that the N- and C-terminal coupling in drug amphiphiles significantly changes the cellular uptake and effective cytotoxicity of the chemotherapeutic agent [73] in analogous drug-sensitive and drug-resistant cell lines. In their study, both Doxorubicin (a small-molecule chemotherapy drug that freely diffuses across cell membranes; abbreviated Dox) and 5-fluorocarboxyfluorescein (5-FAM) were conjugated via a peptide linker known to be cleavable by lysosomal enzymes such as Cathepsin B [27,74] to the N- and C-terminus of a cell-penetrating peptide. Just as entire protein function can depend upon a free and accessible terminal amine or carboxylate group [75], conjugation of a molecule to one of these termini may drastically change its function and overall bioactivity. Indeed, conjugation site in the drug amphiphiles significantly varied each compound's behavior; conjugation at the C-terminal end favored an increase in both internalization and also cytotoxicity of the drug cargo in drug-resistant cancer cell lines [73]. It is believed that the cytotoxic effects induced in the drug-resistant cell lines is due to the inability of Dox to be efficiently flushed from the cell by P-glycoproteins, effusive membrane transporters that are overexpressed in drug-resistant phenotypes [76,77], in its Tat-conjugated form. Therefore, C-terminal Tat conjugation to a chemotherapeutic via a lysosomal protease-cleavable linker does not interfere with the absorption-mediated endocytosis-inducing bioactivity of the terminal amino group [78] and also allows for increased cytosolic retention in drug-resistant cells.

5.4.2 Cleavable Linkers and Auxiliary Segments

As the cellular entry of these drug amphiphile systems becomes more efficient, they must continue to maintain their ability to effectively release bioactive cargo inside their target [79]. To this end, degradable linkers that act as stimuli-activated biosensors [80] that can detect changes in pH [81], reducing agents [82,83], or enzyme concentrations [84] show great potential. To gain further insight into cleavage-activated drug release, studies were done on the previously discussed Tau-conjugated CPT amphiphile, which contains a disulfanyl propyl ester linker that is reducible by glutathione (GSH). Analysis of the distribution of intermediate compounds following the initial degradation of the CPT-disulfide-Tau complex revealed the formation of large aggregations of disulfide dimers that sterically blocked effective cleavage and release of CPT [79]. To better tune the release rate of this drug amphiphile, a self-immolating disulfanyl-ethyl carbonate linker was incorporated [82,83], preventing disulfide bond formation between intermediates and more rapid degradability for more effective drug release. *In vitro* evaluation of this auxiliary addition demonstrates a more therapeutically efficient drug amphiphile [79]. Such a finding places emphasis on the design of the linkage segment when considering the desired function of any nanotherapeutic.

Although the usage of Tat peptides as covalent carriers has become quite advantageous, it is problematic that such cell-penetrating peptides can actually hinder the intracellular uptake of drugs that can freely traverse cell membranes in drug-sensitive phenotypes [73,85]. With respect to its mechanism of cellular entry, Cui et al. have shown through usage of Tat peptides with auxiliary conjugations that electrostatic attraction between the cell-penetrating peptides (CPP) and the cell membrane is not the sole determinant of penetrating efficiency [55,73]. In light of recent studies highlighting the importance of the Tat peptide's hydrophobic section [43] in cellular entry, auxiliary hydrophobic segments can be added to further facilitate cell penetration, as well as reduce differences in cellular uptake between drug sensitive and resistant cell lines alike [85]. These proofs of concepts were demonstrated first with palmitoylated Tat-conjugated fluorescent dyes, and then demonstrated more practically by replacing the hydrocarbon segment with a hydrophobic light-quencher [85]; such a molecule may function as an intracellular nanobeacon that will fluoresce when encountering overexpressed amounts of specific enzymes. It was found that cellular uptake and cytotoxic efficacy were consistent across sensitive and resistant cell phenotypes; this finding holds promise for the even distribution of therapeutics across the heterogeneity of á tumor microenvironment, one of the major barriers for current chemotherapeutics. To this end, cell-penetrating peptides can gain additional functionality through the rational conjugation of a bioactive, hydrophobic auxiliary component.

5.5 CONCLUSION

In this chapter, we have examined the need for nanoscale-based therapies for improved medical regimes, the inherent demands of designing such systems, the power of peptide-based self-assembly to fabricate discrete nanostructures to meet these design standards, and the current promise and success of drug amphiphiles as anticancer agents. Future research will be geared toward increasingly specific targeting, higher drug-loading efficiencies, optimization of self-deliverable hydrogels for local administration, the incorporation of multifunctional drug carriers into delivery systems for dual therapeutic and diagnostic (imaging) purposes, and the safe and cost-effective scale-up of these therapies to meet clinical demand. Indeed, the appearance of such designs will undoubtedly change the face of medicine, offering new solutions for the fight against disease.

REFERENCES

[1] S.E. McNeil, Nanoparticle therapeutics: a personal perspective, WIREs Nanomed Nanobiotechnol 1 (2009) 264–271.

[2] Y. Matsumura, H. Maeda, A new concept for macromolecular therapeutics in cancer chemotherapy: mechanism of tumoritropic accumulation of proteins and the anti-tumor agent smancs, Cancer Res 46 (1986) 6387–6392.

[3] R. Duncan, The dawning era of polymer therapeutics, Nat Rev Drug Discov 2 (2003) 347–360.

[4] Y. Geng, P. Dalhaimer, S. Cai, R. Tsai, M. Tewari, T. Minko, Shape effects of filaments versus spherical particles in flow and drug delivery, et al. Nat Nanotechnol 2 (2007) 249–255.

[5] E. Beniash, J.D. Hartgerink, H. Storrie, J.C. Stendahl, S.I. Stupp, Self-assembling peptide amphiphile nanofiber matrices for cell entrapment, Acta Biomater 1 (2005) 387–397.

[6] S.E. Gratton, P.A. Ropp, P.D. Pohlhaus, J.C. Luft, V.J. Madden, M.E. Napier, The effect of particle design on cellular internalization pathways, et al. Proc Natl Acad Sci USA 105 (2008) 11613–11618.

[7] S.K. Sahoo, V. Labhasetwar, Nanotech approaches to drug delivery and imaging, Drug Discov Today 8 (2003) 1112–1120.

[8] T.M. Allen, Drug delivery systems: entering the mainstream, Science 303 (2004) 1818–1822.

[9] P.A. Netti, S. Roberge, Y. Boucher, L.T. Baxter, R.K. Jain, Effect of transvascular fluid exchange on pressure–flow relationship in tumors: a proposed mechanism for tumor blood flow heterogeneity, Microvasc Res 52 (1996) 27–46.

[10] J.R. Baker, Why I believe nanoparticles are crucial as a carrier for targeted drug delivery, WIREs Nanomed Nanobiotechnol 5 (2013) 423–429.

[11] M.S. Goldberg, S.S. Hook, A.Z. Wang, J.W. Bulte, A.K. Patri, F.M. Uckun, Biotargeted nanomedicines for cancer: six tenets before you begin, et al. Nanomedicine 8 (2013) 299–308.

[12] Q. Sun, M. Radosz, Y. Shen, Challenges in design of translational nanocarriers, J Control Release 164 (2012) 156–169.

[13] D.W. Bartlett, H. Su, I.J. Hildebrandt, W.A. Weber, M.E. Davis, Impact of tumor-specific targeting on the biodistribution and efficacy of siRNA nanoparticles measured by multimodality in vivo imaging, Proc Natl Acad Sci USA 104 (2007) 15549–15554.

[14] D.B. Kirpotin, D.C. Drummond, Y. Shao, M.R. Shalaby, K. Hong, U.B. Nielsen, Antibody targeting of long-circulating lipidic nanoparticles does not increase tumor localization but does increase internalization in animal models, et al. Cancer Res 66 (2006) 6732–6740.

[15] M. Liu, J.M. Fréchet, Designing dendrimers for drug delivery, Pharm Sci Technol Today 2 (1999) 393–401.

[16] E.J. Chung, Y. Chen, R. Morshed, K. Nord, Y. Han, M.L. Wegscheid, Fibrin-binding, peptide amphiphile micelles for targeting glioblastoma, et al. Biomaterials 35 (2014) 1249–1256.

[17] Y. Shiose, H. Kuga, H. Ohki, M. Ikeda, F. Yamashita, M. Hashida, Systematic research of peptide spacers controlling drug release from macromolecular prodrug system, carboxymethyldextran polyalcohol-peptide conjugates, Biconjug Chem 20 (2009) 60–70.

[18] J.M. Meerun Terwogt, G. Groenewegen, D. Pluim, M. Maliepaard, M.M. Tibben, A. Huisman, Phase I and pharmacokinetic study of SPI-77, a liposomal encapsulated dosage form of cisplatin, et al. Cancer Chemother Pharmacol 49 (2002) 201–210.

[19] A. Cabanes, K.E. Briggs, P.C. Gokhale, J.A. Treat, A. Rahman, Comparative in vivo studies with paclitaxel and liposome-encapsulated paclitaxel, Int J Oncol 12 (1998) 1035–1040.

[20] M. Mahmoudi, H. Hoffman, B. Rothen-Rutishauser, A. Petri-Fink, Assessing the in vitro and in vivo toxicity of superparamagnetic iron oxide nanoparticles, Chem Rev 112 (2012) 2323–2338.

[21] G.M. Whitesides, B. Grzybowski, Self-assembly at all scales, Science 295 (2002) 2418–2420.

[22] H. Cui, M.J. Webber, S.I. Stupp, Self-assembly of peptide amphiphiles: from molecules to nanostructures to biomaterials, Biopolymers 94 (2009) 1–18.

[23] G.M. Whitesides, J.P. Mathias, C.T. Seto, Molecular self-assembly and nanochemistry: a chemical strategy for the synthesis of nanostructures, Science 254 (1991) 1312–1319.

[24] A. Trent, R. Marullo, B. Lin, M. Black, M. Tirrell, Structural properties of soluble peptide amphiphile micelles, Soft Matter 7 (2011) 9572–9582.

[25] P. Berndt, G.B. Fields, M. Tirrell, Synthetic lipidation of peptides and amino acids: monolayer structure and properties, J Am Chem Soc 117 (1995) 9515–9522.

[26] T. Gore, Y. Dori, Y. Talmon, M. Tirrell, H. Bianco-Peled, Self-assembly of model collagen peptide amphiphiles, Langmuir 17 (2001) 5352–5360.

[27] T. Etrych, L. Kovar, J. Strohalm, P. Chytil, B. Rihova, K. Ulbrich, Biodegradable star HPMA polymer–drug conjugates: biodegradability, distribution and anti-tumor efficacy, J Control Release 154 (2011) 241–248.

[28] Y. Chau, Y. Luo, A.C.Y. Cheung, Y. Nagai, S.G. Zhang, J.B. Kobler, Incorporation of a matrix metalloproteinase-sensitive substrate into self-assembling peptides – a model for biofunctional scaffolds, et al. Biomaterials 29 (2008) 1713–1719.

[29] K.M. Galler, L. Aulisa, K.R. Regan, R.N. D'Souza, J.D. Hartgerink, Self-assembling multi-domain peptide hydrogels: designed susceptibility to enzymatic cleavage allows enhanced cell migration and spreading, J Am Chem Soc 132 (2010) 3217–3223.

[30] M.C. Giano, D.J. Pochan, J.P. Schneider, Controlled bio-degradation of self-assembling β-hairpin peptide hydrogels by proteolysis with matrix metalloproteinase-13, Biomaterials 32 (2011) 6471–6477.

[31] Y.A. Lin, Y.C. Ou, A.G. Cheetham, H. Cui, Rational design of MMP-degradable peptide-based supramolecular filaments, Biomacromolecules 15 (2014) 1419–1427.

[32] H. Storrie, M.O. Guler, S.N. Abu-Amara, T. Volberg, M. Rao, B. Geiger, Supramolecular crafting of cell adhesion, et al. Biomaterials 28 (2007) 4608–4618.

[33] J.D. Hartgerink, E. Beniash, S.I. Stupp, Peptide-amphiphile nanofibers: a versatile scaffold for the preparation of self-assembling materials, Proc Natl Acad Sci USA 99 (2002) 5133–5138.

[34] P. Zhang, L. Hu, Q. Yin, Z. Zhang, L. Feng, Y. Li, Transferrin-conjugated poly-phosphoester hybrid micelle loading paclitaxel for brain-targeting delivery: synthesis, preparation, and in vivo evaluation, J Control Release 159 (2012) 429–434.

[35] U. Kedar, P. Phutane, S. Shidhaye, V. Kadam, Advances in polymeric micelles for drug delivery and tumor targeting, Nanomedicine 6 (2010) 714–729.

[36] S.M. Chang, J.G. Kuhn, H.I. Robins, S.C. Schold, A.M. Spence, M.S. Berger, A phase II study of paclitaxel in patients with recurrent malignant glioma using different doses depending upon the concomitant use of anticonvulsants, et al. Cancer 91 (2001) 417–422.

[37] K.N. Sugahara, T. Teesalu, P.P. Karmali, V.R. Kotamraju, L. Agemy, D.R. Greenwald, Coadministration of a tumor-penetrating peptide enhances the efficacy of cancer drugs, et al. Science 328 (2010) 1031–1035.

[38] V.P. Torchilin, Targeted pharmaceutical nanocarriers for cancer therapy and imaging, AAPS J 9 (2007) E128–E147.

[39] S. Dagar, A. Krishnadas, I. Rubinstein, M.J. Blend, H. Onyüksel, VIP grafted sterically stabilized liposomes for targeted imaging of breast cancer: in vivo studies, J Control Release 91 (2003) 123–133.

[40] J.P. Richard, K. Melikov, E. Vives, C. Ramos, B. Verbeure, M.J. Gait, Cell-penetrating peptides – a reevaluation of the mechanism of cellular uptake, et al. J Biol Chem 278 (2003) 585–590.

[41] S. Deshayes, M.C. Morris, G. Divita, F. Heitz, Cell-penetrating peptides: tools for intracellular delivery of therapeutics, Cell Mol Life Sci 62 (2005) 1839–1849.

[42] M. Zorko, U. Langel, Cell-penetrating peptides: mechanism and kinetics of cargo delivery, Adv Drug Deliv Rev 57 (2005) 529–545.

[43] P.L. Rodriguez, T. Harada, D.A. Christian, D.A. Pantano, R.K. Tsai, D.E. Discher, Minimal "self" peptides that inhibit phagocytic clearance and enhance delivery of nanoparticles, Science 339 (2013) 2–6.

[44] J.K. Armstrong, G. Hempel, S. Koling, L.S. Chan, T. Fisher, H.J. Meiselman, Antibody against poly(ethylene glycol) adversely affects PEG-asparaginase therapy in acute lympho-blastic leukemia patients, et al. Cancer 110 (2007) 103–111.

[45] A.L. Klibanov, K. Maruyama, A.M. Beckerleg, V.P. Torchilin, L. Huang, Activity of am-phipathic poly(ethylene glycol) 5000 to prolong the circulation time of liposomes depends on the liposome size and is unfavorable for immunoliposome binding to target, Biochim Biophys Acta 1062 (1991) 142–148.

[46] J.P. Photos, L. Bacakova, B. Discher, F.S. Bates, D.E. Discher, Polymer vesicles in vivo: correla-tions with peg molecular weight, J Control Release 90 (2003) 323–334.

[47] M. Black, A. Trent, Y. Kostenko, J.S. Lee, C. Olive, M. Tirrell, Self-assembled peptide amphiphile micelles containing a cytotoxic T-cell epitope promote a protective immune response in vivo, Adv Mater 24 (2012) 3845–3849.

[48] S. Soukasene, D.J. Toft, T.J. Moyer, H. Lu, H. Lee, S.M. Standley, Antitumor activity of peptide amphiphile nanofiber-encapsulated camptothecin, et al. ACS Nano 5 (2011) 9113–9121.

[49] D. Peer, J.M. Karp, S. Hong, O.C. FaroKhzad, R. Margalit, R. Langer, Nanocarriers as an emerging platform for cancer therapy, Nat Nanotechnol 2 (2007) 751–760.

[50] F. Zhao, M.L. Ma, B. Xu, Molecular hydrogels of therapeutic agents, Chem Soc Rev 38 (2009) 883–891.

[51] Y. Geng, P. Dalhaimer, S.S. Cai, R. Tsai, M. Tewari, T. Minko, Shape effects of filaments versus spherical particles in flow and drug delivery, et al. Nat Nanotechnol 2 (2007) 249–255.

[52] P. Decuzzi, B. Godin, T. Tanaka, S.Y. Lee, C. Chiappini, X. Liu, Size and shape effects in the biodistribution of intravascularly injected particles, et al. J Control Release 141 (2010) 320–327.

[53] S.E.A. Gratton, P.A. Ropp, P.D. Pohlhaus, J.C. Luft, V.J. Madden, M.E. Napier, The effect of particle design on cellular internalization pathways, et al. Proc Natl Acad Sci USA 105 (2008) 11613–11618.

[54] T.P. Lodge, Block copolymers: past successes and future challenges, Macromol Chem Phys 204 (2003) 265–273.

[55] P. Zhang, A.G. Cheetham, Y.A. Lin, H. Cui, Self-assembled Tat nanofibers as effective drug carrier and transporter, ACS Nano 7 (2013) 5965–5977.

[56] S.E. Paramonov, H.W. Jun, J.D. Hartgerink, Self-assembly of peptide-amphiphile nanofibers: the roles of hydrogen bonding and amphiphilic packing, J Am Chem Soc 128 (2006) 7291–7298.

[57] S. Soukasene, D.J. Toft, T.J. Moyer, H.M. Lu, H.K. Lee, S.M. Standley, Antitumor activity of peptide amphiphile nanofiber-encapsulated camptothecin, et al. ACS Nano 5 (2011) 9113–9121.

[58] L. Hsu, G.L. Cvetanovich, S.I. Stupp, Peptide amphiphile nanofibers with conjugated polydiacetylene backbones in their core, J Am Chem Soc 130 (2008) 3892–3899.

[59] E.G. Bellomo, M.D. Wyrsta, L. Pakstis, D.J. Pochan, T.J. Deming, Stimuli-responsive polypeptide vesicles by conformation-specific assembly, Nat Mater 3 (2004) 244–248.

[60] E.P. Holowka, V.Z. Sun, D.T. Kamei, T.J. Deming, Polyarginine segments in block copolypeptides drive both vesicular assembly and intracellular delivery, Nat Mater 6 (2007) 52–57.

[61] L.L. Lock, M. LaComb, K. Schwartz, A.G. Cheetham, Y.A. Lin, P. Zhang, Self-assembly of natural and synthetic drug amphiphiles into discrete supramolecular nanostructures, et al. Faraday Discuss 166 (2013) 285–301.

[62] P.S. Low, W.A. Henne, D.D. Doorneweerd, Discovery and development of folic acid based receptor targeting for imaging and therapy of cancer and inflammatory diseases, Acc Chem Res 41 (2008) 120–129.

[63] S. Bonazzi, M.M. Demorais, G. Gottarelli, P. Mariani, G.P. Spada, Self-assembly and liquid–crystal formation of folic acid salts, Angew Chem Int Ed 32 (1993) 248–250.

[64] G. Gottarelli, E. Mezzina, G.F. Spada, F. Carsughi, G. DiNicola, P. Mariani, The self-recognition and self-assembly of folic acid salts in isotropic water solution, et al. Helv Chim Acta 79 (1996) 220–234.

[65] K. Kanie, M. Nischii, T. Yasuda, T. Taki, S. Ujiie, T. Kato, Self-assembly of thermotropic liquid-crystalline folic acid derivatives: hydrogen bonded complexes forming layers and columns, J Mater Chem 11 (2001) 2875–2886.

[66] K.M. Huttunen, H. Raunio, J. Rautio, Prodrugs – from serendipity to rational design, J Pharmacol Rev 63 (2011) 750.

[67] J.A. MacKay, M. Chen, J.R. McDaniel, W. Liu, A.J. Simnick, A. Chilkoti, Self-assembling chimeric polypeptide-doxorubicin conjugate nanoparticles that abolish tumors after a single injection, Nat Mater 8 (2009) 993–999.

[68] H. Dong, C. Dong, W. Xia, Y. Li, T. Ren, Self-assembled, redox-sensitive H-shaped pegylated methotrexate conjugates with high drug-carrying capability for intracellular drug delivery, Med Chem Commun 5 (2014) 147–152.

[69] Y. Zhang, Y. Kuang, Y. Gao, B. Xu, Versatile small molecule motifs for self-assembly in water and formation of biofunctional supramolecular hydrogels, Langmuir 27 (2011) 529–537.

[70] A.G. Cheetham, P. Zhang, Y.A. Lin, L.L. Lock, H. Cui, Supramolecular nanostructures formed by anticancer drug assembly, J Am Chem Soc 135 (2013) 2907–2910.

[71] K.J.C. van Bommel, M.C.A. Stuart, B.L. Feringa, J. van Esch, Two-stage enzyme mediated drug release from LMWG hydrogels, Org Biomol Chem 3 (2005) 2917.

[72] H.A. Behanna, J.J.J.M. Donners, A.C. Gordon, S.I. Stupp, Coassembly of amphiphiles with opposite peptide polarities into nanofibers, J Am Chem Soc 127 (2005) 1193–1200.

[73] P. Zhang, A.G. Cheetham, L.L. Lock, H. Cui, Cellular uptake and cytotoxicity of drug-peptide conjugates regulated by conjugation site, Bioconjug Chem 24 (2013) 604–613.

[74] N. Teich, H. Bodeker, V. Keim, Cathepsin B cleavage of the trypsinogen activation peptide, BMC Gastroenterol 2 (2002) 16.

[75] X.X. Sun, C.C. Wang, The N-terminal sequence (residues 1−65) is essential for dimerization, activities, and peptide binding of *Escherichia coli* DSbC, J Biol Chem 275 (2001) 22743–22749.

[76] A. Persidis, Cancer multidrug resistance, Nat Biotechnol 17 (1999) 94–95.

[77] X.W. Dong, R.J. Mumper, Nanomedicinal strategies to treat multidrug-resistant tumors: current progress, Nanomedicine 5 (2001) 597–615.

[78] D. Missirlis, H. Khant, M. Tirrell, Mechanisms of peptide amphiphile internalization by SJSA-1 cells in vitro, Biochemistry 48 (2009) 3304–3314.

[79] A.G. Cheetham, Y.C. Ou, P. Zhang, H. Cui, Linker-determined drug release mechanism of free camptothecin from self-assembling drug amphiphiles, Chem Commun 50 (2014) 6039.

[80] A.J.M. D'Souza, E.M. Topp, Release from polymeric prodrugs: linkages and their degradation, J Pharm Sci 93 (2004) 1962–1979.

[81] D.Y. Furgeson, M.R. Dreher, A. Chilkoti, Structural optimization of a "smart" doxorubicin-polypeptide conjugate for thermally targeted delivery to solid tumors, J Control Release 110 (2006) 362–369.

[82] W.A. Henne, D.D. Doorneweerd, A.R. Hilgenbrink, S.A. Kularatne, P.S. Low, Synthesis and activity of a folate peptide camptothecin prodrug, Bioorg Med Chem Lett 16 (2006) 5350–5355.

[83] E.A. Dubikovskaya, S.H. Thorne, T.H. Pillow, C.H. Contag, P.A. Wender, Overcoming multidrug resistance of small-molecule therapeutics through conjugation with releasable octaarginine transporters, Proc Natl Acad Sci USA 105 (2008) 12128–12133.

[84] D. Gabriel, M.F. Zuluaga, H. van den Bergh, R. Gurny, N. Lange, It is all about proteases: from drug delivery to in vivo imaging and photomedicine, Curr Med Chem 18 (2011) 1785–1805.

[85] P. Zhang, L.L. Lock, A.G. Cheetham, H. Cui, Enhanced cellular entry and efficacy of Tat conjugates by rational design of the auxiliary segment, Mol Pharm 11 (2014) 964–973.

Printed and bound by CPI Group (UK) Ltd, Croydon, CR0 4YY

03/10/2024

01040423-0010